Forecasting with Maximum Entropy

The interface between physics, biology, economics and information theory

Online at: https://doi.org/10.1088/978-0-7503-3931-5

Forecasting with Maximum Entropy

The interface between physics, biology, economics and information theory

Hugo Fort
Department of Physics, Republic University, Montevideo, Uruguay

IOP Publishing, Bristol, UK

ISBN 978-0-7503-3931-5 (ebook)
ISBN 978-0-7503-3929-2 (print)
ISBN 978-0-7503-3932-2 (myPrint)
ISBN 978-0-7503-3930-8 (mobi)

DOI 10.1088/978-0-7503-3931-5

Version: 20221101

IOP ebooks

British Library Cataloguing-in-Publication Data: A catalogue record for this book is available from the British Library.

Published by IOP Publishing, wholly owned by The Institute of Physics, London

IOP Publishing, No.2 The Distillery, Glassfields, Avon Street, Bristol, BS2 0GR, UK

US Office: IOP Publishing, Inc., 190 North Independence Mall West, Suite 601, Philadelphia, PA 19106, USA

To the memory of my parents.

Contents

Preface

'But ignorance of the different causes involved in the production of events, as well as their complexity, taken together with the imperfection of analysis, prevents our reaching the same certainty about the vast majority of phenomena. Thus there are things that are uncertain for us, things more or less probable, and we seek to compensate for the impossibility of knowing them by determining their different degrees of likelihood. So it was that we owe to the weakness of the human mind one of the most delicate and ingenious of mathematical theories, the science of chance or probability.'

— Pierre-Simon Laplace (1776)

I began thinking about this book ten years ago when, after working in apparently different problems in distinct fields, I noticed profound similarities between them and the methods which are commonly used to approach these problems. But, at same time, I also noticed there was a lack of communication between disciplines that precludes applying methods which demonstrated its power in one discipline to formally similar problems in other areas. In fact researchers tend to work in one or another sub-field, often in ignorance of parallel developments in other disciplines. More importantly, the innovative and transformative science that leads to breakthroughs often happens at the intersections of disciplines, and this isolation of disciplines from one another is a barrier for integrating work across disciplinary boundaries. Additionally, global problems such as the rapid loss of natural resources or the impact on climate resulting from human activities cannot be addressed by scientists from single disciplines. Instead, over the past decades the conduct of scientific research has shifted away from single principal investigators working in departments aligned with traditional scientific disciplines to collaborative, problem-based, interdisciplinary teams that span institutional boundaries.

Thereby, the purpose of this book is threefold. First, it aims at providing a unifying framework, based on information entropy and its maximization, to connect fundamental issues in biology (mainly ecology and evolution), quantitative and financial economics, and statistical physics. A dialogue between disciplines is the main condition for interdisciplinary research aimed at solving or mitigating global problems and at transformative advances in science. Therefore, in addition to advances in the underlying disciplines, such research challenges require promoting convergence, i.e., the merging of ideas, approaches and tools from widely diverse fields of knowledge to accelerate innovation and discovery. Second, and closely connected with the first objective, is to share problem-solving strategies. That is, this book attempts to bring together practitioners from the above disciplines who are interested in applying methods from other fields to approach their own research problems. The third objective is to provide a forecasting method for important practical problems in different disciplines. To achieve this goal we combine analytical treatments, to obtain exact closed formulas condensing basic relationships among the relevant variables for sensible stylized or minimal modelling, with numerical calculations, for more realistic descriptions of systems and empirical data. This is done always ensuring quality of presentation as well as content. The wide-ranging scope of the problems considered will not sacrifice the depth to which they are treated. Hopefully, it will show the usefulness of physics inspired modelling to solve practical problems in biology, economics and environmental science.

The common thread throughout the book is how the flux of information controls as well as serves to predict the dynamics of complex systems. Information emerges as a key concept across biological systems, markets and physical systems. Hence, to provide an integrative perspective for the interface between physics, biology and economics we rely on information theory (IT), which, since its formulation by Claude Shannon in the late 1940s, has emerged as a key conceptual framework across different scientific areas. Along the chapters we will proceed like putting together a puzzle, one piece per chapter[1], and building conceptual bridges between the new piece with the other pieces that have already been assembled.

Therefore, we start in chapter 1 by reviewing a main concept of IT, namely *information entropy*, introduced by Shannon to quantify the idea of *missing information* (MI). In the late 1950s, E T Jaynes introduced their principle of maximum entropy (MaxEnt) which clarified the precise connection of Shannon's entropy to the statistical thermodynamics entropy. Furthermore, this principle, has demonstrated to be a powerful general method to make the least biased inferences compatible with available data in a variety of realms for dealing with uncertainty within a decision analytic framework. These decision analysis problems are quite common and in fact are fundamental in operations research practice, in designing financial strategies or in environmental management.

This relationship between information entropy and thermodynamics entropy is discussed with greater depth in chapter 2. We start by introducing the basics of statistical mechanics. Statistical mechanics founded by Maxwell, Boltzmann and Gibbs in the mid-19th century in an effort to explain the phenomenological laws of thermodynamics from the more fundamental Newton's laws. Then we show that it is possible to define the amount of information obtained from a certain experiment, and to measure it in a precise way. As a matter of fact, information can be actually regarded as a physical quantity that can profoundly affect the way that matter behaves in everyday life. If we focus on open systems interacting with their environment, like ecosystems or markets, information entropy plays a central role. Interestingly, at the 'hardware level', where entropy is synonym of disorder, such systems rely on the exportation of entropy to the environment to achieve a state of low entropy that maintains them in a steady self-organized state that allows them to properly work. This is a state of *minimum entropy*, far from thermodynamic equilibrium (which is a state of maximum entropy). On the other hand, at the 'software level', *maximum entropy* is a keyword to approach the network-like

[1] At the end of each chapter we will show the puzzle with the added piece.

complex organizational structures displayed by either biological or economic systems. In fact, Jaynes' MaxEnt is a general method to draw inferences from incomplete information of these complex systems by assuming maximal ignorance about the unknown degrees of freedom. We show that the MaxEnt Principle thereby provides an additional methodological link between biology, economics and physics.

Chapter 3 begins with an interesting mathematical connection between biology, more precisely ecology, and physics by noticing that the linear Lotka–Volterra generalized equations (LLVGE) are similar to the equations used in chemical kinetics. Indeed, when proposing these LLVGE, Lotka sought to develop for biological systems involving different interacting species a statistical mechanics approach similar to the one Boltzmann and Gibbs had developed to deal with the kinetic theory of gases and related problems. Since then these equations have been the basis of classical population ecology and also widely used in theoretical biology.

In turn, in chapter 4, we present formal analogies of economics, first with physics, and next with biology. We start by discussing the remarkable mutual influence between physics and economics along the 19th century. Indeed this was a two-way street. On the one hand we have the neoclassical theory tradition of formulating *economics as physics*, inspired in the mechanics metaphor, which forms the basis of mainstream economics. On the other hand and less widely known, economics contributed to physics with many of the metaphorical and experimental guidelines needed to translate thermodynamic phenomena into quantifiable terms. However, an important objection to neoclassical theory is that it relies almost exclusively on the axiomatic-deductive approach while empirical evidence contradicting neoclassical assumptions is generally ignored. Another critique is about rational choice theory which assumes that people always make optimal decisions that provide them with the greatest benefit and satisfaction. This would happen in an ideal world. In the real world, people are instead often moved by emotions and external factors. This brings us to *economics as biology*, i.e., to the so-called evolutionary economics in which evolutionary ideas are used to explain the nature of the processes of innovation and the institutions supporting them. An important tool of evolutionary economics is the replicator dynamics equation (RDE), which describes the evolution of markets in terms of a *payoff matrix*. The element *ij* of this matrix is the 'payoff' received by firm number *i* from the interaction with firm *j*. Interestingly, the RDE is mathematically equivalent to the ecological LLVGE of chapter 3, and its payoff matrix is basically the same as the Lotka–Volterra generalized matrix quantifying the strength of interactions between species. This builds a link connecting the new piece of evolutionary economics with the piece to the puzzle that we added in the previous chapter, namely population dynamics.

Thus, a central object to model the dynamics of complex systems like ecosystems or markets is the matrix of pairwise interactions between entities (coexisting species and firms in the case of ecosystems and markets, respectively). The standard recipe to estimate this matrix in artificially assembled communities (e.g., in agricultural sciences or microbial experiments) is to isolate the species into monoculture and biculture experiments to gauge the effects of pairwise interactions with respect to the species alone. Nevertheless, this procedure of breaking the system into pieces is often not applicable in more complex communities, like natural ecosystems, let alone markets, where firms rely on other firms to be their customers and depend on complex market chains. We therefore have to devise alternative procedures to estimate this matrix by more indirect methods. A way of inferring an effective interaction matrix M between the constituting entities of a complex system, without isolating them, is through the MaxEnt. This is the topic of chapter 5 which is the central piece of the puzzle or keystone of this book. As we have seen in chapters 1 and 2, the basis of the MaxEnt method is to work with known constraints on the system written as average values. Thus, in order to obtain a good candidate for the interaction matrix through MaxEnt, a key element is choosing the right constraints. In particular, the choice of the covariance matrix, which connects pairs of species, as one of the constraints is crucial to obtain a set of pairwise interaction coefficients. The models that result when we maximize the Shannon information entropy imposing that the means and covariances of entities (species, firms, etc) are known as pairwise maximum-entropy (PME) models.

There are diverse examples in which such PME models have been used to analyze data associated with biological problems in ecology and evolution. In chapter 6 we present an application of PME modelling to conservation ecology. A challenge for conservation biologists and agencies working to sustain the ecosystem services is how to reduce the vulnerability of biodiversity of an ecological community to several factors like climate change or resource exploitation. Actually, there is a clear need for early warning indicators of species loss generated from empirical data before these losses occur. With this goal in mind we consider the tree community of the long-term 50-hectare plot on Barro Colorado Island (BCI), Panama, which is one of the most intensively studied in the world. This plot was established in 1981 and fully censused in 1982, and then every five years. This extensive dataset reveals that some tree species suffered steep population declines. We propose an early warning indicator of such tree population crashes based on PME and test it against the BCI dataset. Specifically, we compute the effective interaction matrices, M, among species for the eight censuses available. For each species i and each census c, the absolute value of the intraspecific competition coefficients $M_{ii}(c)$ are much larger than those of the interspecific interaction coefficients $M_{ij}(c)$ with $i \neq j$. Furthermore, for those tree species that suffered steep population declines (of at least 50%), across the eight tree censuses, the drop of M_{ii} is always steeper and occurs before the drop of the corresponding species abundance N_i. Indeed, such sharp declines in M_{ii} occur between 5 and 15 years in advance of comparable declines for N_i, and thus they serve as early warnings of impending population busts.

The final chapter 7 addresses the so-called 'Marshall's problem', namely integrating physics and biology into economics in a way that both contains its orthodox core as well as provides a transdisciplinary perspective on economic complexity (Cassata and Marchionatti 2011). This is a main theme of this book which completes the puzzle. We propose a method based on the replicator dynamics (discussed in chapter 4) as an equation to model natural selection in markets. A main difficulty is how to obtain the payoff matrix connecting the pairwise effects between interacting market entities. We thus use the PME procedure to estimate the payoff matrix. The resulting method is called replicator dynamics pairwise maximum entropy (RDPME). We test the forecasting performance of RDPME using daily market values from 2014 to 2019 of America's top revenue companies. The main finding is that the RDPME method outperforms the stochastic benchmark of the geometric random walk in predicting empirical shares for most of the companies along most choices of validation periods.

Indeed, there are also important differences between markets, and physical or even biological systems that are worth mentioning. An important problem is that in economics, unlike natural science, controlled experiments cannot be conducted. As Samuelson and Nordhaus argue in their classical textbook, Economics (1985):

> 'Economists (unfortunately) cannot perform the controlled experiments of chemists or biologists because they cannot control other important factors. Like astronomers or meteorologists they generally must be largely content to observe.'

There are different reasons precluding controlled experiments in economics. One is that in economics—just like in any other social science—it is very difficult to claim that an event is the result of one cause only. Instead, often there are many causes and vice versa, one cause can generate several consequences. In addition, frequently it is not possible to change an independent variable intentionally for either ethical or practical reasons. Nevertheless, in spite of the above limitations, the application of experimental methods in economics has grown considerably over the last few decades and it has given researchers in the field access to previously unobtainable data that can be used to validate or refute theoretical propositions.

Another important difference is that markets, like any other systems involving human actions, decisions, reactions are by their very nature *adaptive*. This means that the very act of predicting their behavior becomes a factor that adds information, which in turn alters the system and influences this behavior. This is the self-fulfilling prophecy effect, i.e., when a prediction directly or indirectly causes itself to become a reality, by the very terms of the prophecy itself, due to positive feedback between the belief and the behavior. For instance, predictions of economic growth influence the investment decisions of firms, which, in turn, affect the rate of economic growth. Another classic example of a self-fulfilling prophecy is the bank failures during the Great Depression that took place mostly during the 1930s, when even banks on strong financial footing sometimes were driven to insolvency by bank runs. Often, if a false rumor started that the bank was incapable of covering its deposits, a panic ensued, and depositors wanted to withdraw their money all at once before the bank's cash ran out. When the bank could not cover all the withdrawals, it actually did become insolvent. Thus, an originally false belief led to its own fulfillment (Britannica 2022). The Israeli historian Yuval Noah Harari adopts a radical stance regarding predictability of markets. He claims that markets, like history, are what he calls 'level two chaotic systems' (Harari 2014). Level one is chaos that does not respond to predictions, for example, weather is a level one chaos system because rain or snow will still happen unaffected by a weather forecast. On the other hand, level two chaos is chaos that reacts to predictions about it, and therefore can never be predicted accurately. In his own words:

> 'What will happen if we develop a computer program that forecasts with 100 per cent accuracy the price of oil tomorrow? The price of oil will immediately react to the forecast, which would consequently fail to materialise. If the current price of oil is $90 a barrel, and the infallible computer program predicts that tomorrow it will be $100, traders will rush to buy oil so that they can profit from the predicted price rise. As a result, the price will shoot up to $100 a barrel today rather than tomorrow. Then what will happen tomorrow? Nobody knows.'

Harari certainly raises an interesting point. The viewpoint in this book is that feedback effects and reactivity to predictions of economic systems don't make prediction impossible; they do make it harder than for non-adaptive systems. Systems like markets, in addition to their reactivity to predictions, include two other fundamental sources of unpredictability. The first one, which is shared with ordinary systems non-reactive systems, is the impossibility of truly isolating a system from its environment Let us assume for a moment that we have complete information about a real (and thus by definition complex) system. We still have the rest of the world which surrounds it. The information we have on this environment is always far from complete. This lack of knowledge introduces a probability distribution for the different possible states or outcomes for the combination of the system and its environment and is discussed in chapter 2. The other source of uncertainty is that the human behavior often deviates from rationality. Thus a market is a creature of moods as much as fundamentals. Emotions like fear, risk aversion or collective euphoria drive markets no less than fundamentals but their effects are much harder to weight. Forecasting is inherently uncertain to some extent either for reactive or non-reactive systems. The problem with the example of forecasting oil prices is that no computer program can accurately incorporate the impact of events like the recent invasion of Ukraine by Russia on the oil market. This unexpected event in the system's environment triggered a sequence of even more unexpected events, sometimes amplified by fear, leading to wild variations in the crude price. During the month after the beginning of this invasion predicting the oil market was possible only with large error bars. Consequently, it seems very likely that either for non-adaptive or adaptive systems we will have to content ourselves with computing probabilities for different outcomes. Laplace already warned us of this limitation in the quotation at the beginning.

Acknowledgements

I want to thank several people. First of all Jorge Pullin, who thoroughly went over drafts of the chapters of this manuscript and corrected many typographical and other errors, also pointing out where more explanation was required, and details I had missed. I am also indebted to Malte Faber for his suggestions about chapter 4.

I am grateful to several co-authors of papers cited in this volume: Raul Donangelo, Clive Emary, Ariel Fernández, Tomas Grigera, Pablo Inchausti, Muhittin Mungan, Marten Scheffer, Angel Segura and Egbert van Nes. I have collaborated with them over the years on different topics of mathematical modelling and ecology. Much of the material of this book corresponds to research developed in collaboration with them.

I am indebted to Edgardo Garcia-Alvarez and Fredy R Zypman for interesting discussions.

It has been a pleasure working with the Institute of Physics (IOP) team; particularly with *Ashley Gasque* (Senior Commissioning Editor), *John Navas* (Senior Commissioning Manager) and *Emily Tapp* (Commissioning Editor). All of them provided valuable help and assistance.

Finally I thank my wife Silvia and my sons, José and Rodrigo, for they patience and support during the writing of this book.

Punta del Este, Maldonado, Uruguay
Hugo Fort

Author biography

Hugo Fort

Dr Hugo Fort is Professor at the Physics Department of the Faculty of Sciences of the Republic University (Montevideo, Uruguay) and Head of the Complex System Group. After earning his PhD in Physics from the Autonomous University of Barcelona in 1994 he conducted research on quantum field theory. Since 2001 his scientific interests evolved from theoretical physics to complex systems and mathematical modelling applied to problems in biology, with focus in ecology and evolution.

A main goal of his research is to develop quantitative methods and tools for a wide variety of practical problems in fields ranging from agroeconomics (crop mixtures for overyielding, scientific or precision livestock production) to environmental sciences (early warnings of catastrophic shifts, forecasting biodiversity dynamics), real-time evolution (RNA viral dynamics, bacterial experiments) and econometrics.

Professor Fort is currently involved in several international research collaborations pursuing used-inspired basic science to problems of production optimization and conservation. A central aim is to connect problems of ecology, evolution and economics with well-studied phenomena in physics to gain deeper insight into these problems and to get access to alternative powerful computational tools.

Author of over a hundred articles in scientific journals and book chapters in diverse fields as agriculture sciences, applied mathematics, biology, ecology, physics and social sciences modelling, Professor Fort has taught several courses in mathematical modelling, complex systems, non-linear dynamics and statistical physics. He has also been collaborating in different projects with national agencies as a senior scientific consultant.

He is currently working on:

1. Explaining the diversity of different communities in terms of interspecific interactions.
2. Networks and the link between network architecture and community stability and performance.
3. Game theoretical modelling as a new approach to niche construction and niche engineering.
4. Early warnings of catastrophic shifts (desertification, eutrophication, shifts in tropical forests).
5. Evolution of microbes in real time (RNA viral dynamics, bacterial experiments).
6. Understanding the dynamics of financial markets and developing evolutionary algorithms for market forecasting.

List of acronyms([*])

ASS: *Alternative stable states*
CA: Cellular automaton
EMH: Efficient-market hypothesis
GICS: Global industry classification standard
grw: Geometric random walk
IT: Information theory
LLVGE: Linear Lotka–Volterra generalized equations
MAE: Mean absolute error
MAPE: Mean absolute percentage error
MaxEnt: Maximum entropy
MF: Mean-field
MI: Missing information
MSE: Mean square error
NYSE: New York Stock Exchange
ODE: *Ordinary differential equations*
PME: Pairwise maximum-entropy
RD: Replicator dynamics
RDE: Replicator dynamics equation
RDPME: Replicator dynamics pairwise maximum-entropy
RMPME: Replicator–mutator pairwise maximum-entropy
RMAE: Relative mean absolute error
RWH: Random walk hypothesis
SAD: *Species-abundance distribution*
SAR: *Species–area relationship*

References

Britannica 2022 https://britannica.com/topic/self-fulfilling-prophecy (accessed 4 March 2022)

Cassata F and Marchionatti R 2011 A transdisciplinary perspective on economic complexity. Marshall's problem revisited *J. Econ. Behav. Organ.* **80** 122–36

Harari Y N 2014 *Sapiens: A Brief History of Humankind* (London: Harvill Secker)

Laplace P S 1776 'Recherches, 1°, sur l'Intégration des Équations Différentielles aux Différences Finies, et sur leur Usage dans la Théorie des Hasards' (1773, published 1776) *Oeuvres complètes de Laplace, 14 Vols. (1843–1912)* vol **8** pp 144–5 trans. Charles Coulston Gillispie, Pierre-Simon Laplace 1749–827: A Life in Exact Science (1997), 26

Samuelson A P and Nordhaus D W 1985 *Economics* (New York: McGraw-Hill)

[*] Italic words refer to entries of the glossary.

Chapter 1

Entropy as missing information: from Shannon's information theory to Jaynes' maximum entropy principle

'Gain in entropy always means loss of information and nothing more'.

—G N Lewis (1930)

'Everybody knows that the dice are loaded/Everybody rolls with their fingers crossed'.

—Leonard Cohen (1988)

- We start by reviewing the central role played by information in modern science and everyday life. We also discuss how the processing of information provides a link between biology and economics with physics.
- Next we review the earliest attempts to quantify information content, developed in the 1920s by Henry Nyquist and later extended by Ralph Hartley, which allowed estimating the capacity of a physical system to transmit information in terms of a logarithmic expression. In particular, the contribution by Hartley was that we should eliminate any psychological factor from a truly quantitative measure of information, instead this measure must be based on physical considerations alone.
- The works of Nyquist and Hartley in the 1920s were influential for Shannon to develop their information theory in the late 1940s. A central concept of this theory was that of information entropy. We provide examples of information entropy and its equivalence to missing information (MI) through some simple worked examples.
- We conclude with Jaynes' principle of maximum entropy (MaxEnt) introduced in the late 1950s, which clarified the precise connection of Shannon's entropy to the statistical thermodynamics entropy. Jaynes also realized that MaxEnt is a general method to draw inferences from incomplete information

doi:10.1088/978-0-7503-3931-5ch1

by assuming maximal ignorance about the unknown degrees of freedom. Since then, information theory has then been applied in such diverse areas as complexity theory, networking analysis, community ecology, finances and mathematical statistics. We discuss the MaxEnt method and provide some simple examples of its application.

1.1 Information and its processing in biology, economics and physics

The exact meaning of the term 'information' varies across different contexts and different disciplines like physics, mathematics, logic, biology or economics. In more colloquial speech information is currently used to denote any amount of data, code or text that is stored, sent, received or manipulated in any medium. Everything we know about the world is based on information we received or gathered. The concept of information has thus played a central role since the ancient Greeks. In fact, we can still recognize the root 'form' in the word *in-formation* (Capurro and Hjørland 2003). According to Plato's Theory of Forms, forms are the real essences of all things. Every object or quality in reality has a form: horses, triangles, mountains, love, and goodness. Plato supposed that what 'really' exist are forms and that the phenomena were mere shadows mimicking these forms. Consequently, information has become a central issue in almost all of the sciences and humanities. The first thing a scientist has to do before he/she can formulate a theory is to gather information. The application possibilities of information theory are abundant. Datamining and the handling of extremely large data sets seems to be an essential for almost every empirical discipline in the 21st century (Adriaans 2020).

How the information flows and is processed throughout different biological or economic systems—be it our brain, society, an organization, an ecosystem, a market—is crucial to understanding their behavior. Today, information theory—the scientific study of the quantification, storage, and communication of information—is used in a wide variety of fields. In particular, biology and economics are two major examples of fields where people are starting to apply information theoretic tools.

In the case of biology, information processing is the essential thing that distinguishes the living from the non-living. Life is based on signals, codes, transcription, and translation. Indeed, the *Central Dogma of Molecular Biology* is an explanation of the flow of genetic information within a biological system. Roughly, it is often stated that 'DNA makes RNA (transcription stage), and RNA makes protein (translation stage)' (Crick 1958, Leavitt 2004)[1]. This states that once 'information' has passed into protein it cannot get out again. Information means here the precise determination of sequence, either of bases in the nucleic acid or of amino acid residues in the protein.

[1] However, the 'Central Dogma' has had to be revised a bit. It turns out that it is possible to go back from RNA to DNA, and that RNA can also make copies of itself. It is still not possible to go from proteins back to RNA or DNA, and no known mechanism has yet been demonstrated for proteins making copies of themselves.

Markets, in turn, have also been considered as information-processing mechanisms between economic agents in which coherence emerges through the independent actions of large numbers of individuals (Hayek 1937, 1945) for finding and communicating price information (Hurwicz 1973). Additionally, information theoretic tools are useful to model long-term behavior of markets and strategies for maximizing performance in such markets. For an interesting review and synthesis of the uses of information and entropy in econometrics we refer the reader to the book of Golan (Golan 2006). This information-processing viewpoint will be of great help for developing in chapter 4 a parallelism between economics and biology, through the evolutionary economics approach.

To complete the interface between biology, economics and information theory with physics let us analyze the connection of information-processing systems, either biological or economic, with physics. This link can be established in different ways or levels. Let us briefly discuss three of them of particular interest to us.

First, all the above information-processing is made possible by a flow of matter and energy with the environment which reduces the entropy of the system. This exportation of entropy is indispensable for open systems to achieve a state of low entropy that maintain themselves far from thermodynamic equilibrium (which is a state of maximum entropy) (Nicolis and Prigogine 1977). In this way they are able to maintain a steady self-organized state (of minimum entropy) in time that allows them to properly work. This is the case either of living creatures (Camazine *et al* 2003) or of organizations like markets (Jutterström 2018).

Second, both biological and economic systems often display network-like complex structures which are similar to the ones found in the physics of complex systems. For example, in addition to the transmission of genetic information, genes participate also in more complex processes like switching other genes on or off using chemical messengers, often leading to complex networks amenable to analyze through the techniques of statistical physics (Zaman 2009). Furthermore, these gene regulatory networks are chemical circuits that resemble electronic or computing components, sometimes constituting modules or gates that enact logical operations (Alon 2006). Statistical physics has also long been concerned with the appearance of universal features in different subfields of economics. For instance, financial markets represent one of the most complex many-body problems, as each interacting body is a sentient being whose behavior responds to several collective signals and particular feelings and beliefs. Indeed, physicists studying financial markets have found that they exhibit statistical structures akin to those in more physical systems (see for instance Voit 2003).

Last but not least, as we will discuss with more detail in chapter 2, the information can be directly connected with the thermodynamic entropy. That is, it is possible to define the amount of information obtained from a certain experiment, and to measure it in a precise way. We only need to know the uncertainty before and after the observation. If the final uncertainty is very small (very accurate measurement) the information obtained is very large. Whenever an experiment is performed in the laboratory, it is paid by an increase of entropy, and a generalized Carnot principle states that the price paid in increase of entropy must always be larger than the amount of information gained (Brillouin 1962). As we will see, this enables one to solve the paradox of *Maxwell's demon* (Maxwell 1871). As a matter of fact, the lesson of Maxwell's demon is that information is actually a physical quantity that can profoundly affect the way that matter behaves (Davies 2020).

1.2 Uncertainty in communication systems: Shannon entropy

1.2.1 First attempts to quantify information content

We have noticed at the beginning of this chapter that the word 'information' has several meanings. However, all these notions share two central properties:

Information is extensive. This means that the combination of two independent datasets contains more information than each separate dataset. This extensiveness of the information is a synonym of the property of *additivity*.

Information reduces uncertainty. The amount of information we get is proportional to the amount by which it reduces our uncertainty.

We can trace the birth of quantitative theories of information to the 1924 paper from Nyquist in the *Bell System Technical Journal* entitled, 'Certain factors affecting telegraph speed'. Nyquist derived there a formula for the maximum speed W of 'transmission of intelligence' over a telegraph. By the speed of transmission of intelligence he meant the number of characters, representing different letters, figures, etc, which can be transmitted in a given length of time assuming that the circuit transmits a given number of signal elements per unit time. Nyquist's formula is in terms of the number of signal levels m and the bandwidth of the system B (in Hz).

Specifically, Nyquist tells us that we can transmit data at a rate of up to:

$$W = B \log m \tag{1.1}$$

(see box 1.1. for a derivation of this formula). W is an upper bound because a noise-free channel is assumed.

Even though most of Nyquist's paper considers the engineering aspects of communication systems, he was beginning to understand the need to abstract away the actual content of the signal from the information carried within the message. Both, his concept of intelligence and the logarithmic rule that governs the maximum amount of intelligence that can be sent across a telegraph wire coincide with the concept of information and its entropy later introduced by Shannon in 1948 (discussed in the next subsection).

The next important landmark in the foundations of information theory was provided by the American electronics researcher Ralph Vinton Lyon Hartley. He invented the Hartley oscillator and the Hartley transform (an integral transform closely related to the Fourier transform). Hartley was the first one who, in his 1928 paper 'Transmission of information', used the word 'information' to describe the communication through diverse systems including telegraphy, telephony, picture

Box 1.1. Derivation of Nyquist's formula for the transmission of intelligence over telegraphs

Let us assume a code whose characters are all of the same duration. If n is the number of signal elements per character, then the total number of characters which can be construed equals $m \times m \times m$ (n times), i.e., m^n.

In order that two such systems should be equivalent, the total number of characters that can be distinguished should be the same. This implies that

$$m^n = \text{const.} \quad \text{(i)}$$

And taking logarithm at both sides this equation may also be written as:

$$n \log m = \text{const.} \quad \text{(ii)}$$

Thus the number of characters per unit of time which can be transmitted over a circuit is directly proportional to the line speed s and inversely proportional to the number of signal elements per character. Hence, we may write

$$W = s/n. \quad \text{(iii)}$$

Substituting (ii) into (iii) we get:

$$W = s \log m/\text{const.} \quad \text{(iv)}$$

And, since s is proportional to the bandwidth of the system B, equation (iv) may also be written as:

$$W = B \log m. \quad \text{(v)}$$

transmission and television over both wire and radio paths. Hartley did not like that Nyquist's concept of intelligence was still plagued by psychological aspects. So, his main goal was 'to set up a quantitative measure, based on physical as contrasted with psychological considerations', whereby the capacities of the above diverse systems to transmit information may be compared. In doing this Hartley provided a stricter definition for what had until that point been a vague term.

Hartley started by defining the transmission of information—which could be words, sound or anything else—as a sequence of n symbols chosen from an alphabet with S symbols. This allowed representing information as a quantity that determined the ability of the receiver of a transmission to determine if a particular sequence was intended by the sender, regardless of the content of the message. The idea is that in any given communication the sender selects a particular symbol and causes the attention of the receiver to be directed to that particular symbol. By successive selections a sequence of symbols is brought to the receiver's attention. At each selection there are eliminated all of the other symbols which might have been chosen. As the selections proceed, more and more possible symbol sequences are eliminated, and we say that the information becomes more precise (i.e., by property II—information reduces uncertainty). The following example by Hartley (1928) is very illuminating:

'...in the sentence, ≪Apples are red,≫ the first word eliminates other kinds of fruit and all other objects in general. The second directs attention to some property or condition of apples, and the third eliminates other possible colors. It does not, however, eliminate possibilities regarding the size of apples, and this further information may be conveyed by subsequent selections.'

At each selection there are available S possible symbols. Two successive selections make possible S^2 different permutations or symbol sequences. Similarly n selections make possible S^n different sequences.

Then Hartley proposed a measure of uncertainty when we sample an element from a finite set S with the uniform distribution, called the *Hartley function* (Hartley 1928) in his honor. To derive this function he reasoned that, in order for a measure of information to be of practical engineering value, it should be of such a nature that the information is proportional to the number of selections n. In addition, this amount of information also depends on the number of possible symbols S as well. That is, a symbol from a two-symbol alphabet (0–1) caries less information than a letter of the English alphabet. Thus we can write

$$H = K(S)n, \tag{1.2}$$

where $K(S)$ is a constant which depends on the number S of symbols available at each selection. Since taking logarithm (in an arbitrary base b) of the number of possible symbol different sequences produces:

$$\log_b S^n = n \log_b S,$$

comparing with equation (1.2) Hartley found it was natural to take $K(S) = \log_b S$, which thus implies that

$$H = \log_b S^n. \tag{1.3}$$

That is, Hartley's measure of information is the logarithm of the number of possible symbol sequences. If the base b of the logarithm is 2, then the unit of uncertainty is the *bit*. If it is the natural logarithm, then the unit is the *nat*. Hartley used a base-ten logarithm, and with this base, the unit of information is called the *dit*.

Remark: Hartley derived his formula based on the two general properties of information we mentioned at the beginning of this section—(I) extensiveness and (II) information reduces uncertainty—plus the assumption that each symbol had equal probability, which is behind the arbitrary choice of setting the amount of information proportional to the number of selections (and so choose the factor of proportionality as to make equal amounts of information correspond to equal numbers of possible sequences).

In fact, Hartley's measure of information contained no notion about the communication of messages of ***unequal probability*** (this aspect was addressed by Shannon).

1.2.2 Shannon's information entropy

The theory of information developed by Claude Shannon in the late 1940s (Shannon 1948) has had a great impact on many fields of science and technology. Before his information theory, the different types of communication had all been transmitted using entirely different mediums, and communication was strictly an engineering discipline, with little scientific theory to back it up. That is, the communication devices of the time (telegraph, telephone wireless telegraph, AM radio, TV, etc) were diverse not just in the media used to deliver the message, but also in the methods used to code and transfer messages from one point to another. Separate fields emerged to deal with the problems associated with each medium, each with their own set of tools and methodologies. For example, it would have been inconceivable to an engineer that one would be able to send video over a phone line, as is commonplace today with the advent of the modem. This is because engineers treated both of these as separate entities and did not see the connection in the transmission of information a concept that would cross the boundaries of these disparate fields and bind them together (Chiu *et al* 2001).

Shannon theory is based on two main ideas. The first one, inspired by the pioneering works by Harry Nyquist and Ralph Hartley, is that information can be treated as a physical quantity which then can be precisely and objectively measured. The second idea can be traced from the method Shannon developed for using Boolean logic to represent electrical circuits. This led him to prove that all communication, as diverse as radio waves, text, pictures and telephone signaling, can be programmed in binary numbers. Such a general inclusive approach to communication, based on these two pillars, was groundbreaking and laid out the foundations of the *information age*, i.e., of an economy primarily based upon *information technology* (Castells 1996). For instance, the development of communication systems and networks has benefited greatly from Shannon's work. Shannon's work on data storage, compression and transmission also paved the way for such inventions as CDs, DVDs, MP3s and JPEGs, among others.

Shannon was originally interested in a very practical problem, namely the capacities of telecommunication lines to transmit information (Shannon 1948). By Shannon's own words: 'The fundamental problem of communication is that of reproducing at one point either exactly or approximately a message selected at another point.' It is worth remarking that Shannon, in the same spirit of Hartley, used the word information, in his theory in a special sense in which all semantic aspects of communication are irrelevant, i.e., information must not be confused with meaning. In fact, two messages, one of which is heavily loaded with meaning and the other of which is pure nonsense, can be exactly equivalent, from the present viewpoint, as regards information (Shannon and Weaver 1949).

Let us thus begin with the first model of the communication process developed by Shannon, which was originally proposed for the use of mechanical messaging for the Bell Telephone Company. It describes communication exchange in terms of five parts (Shannon 1948):

1. An *information source* which produces a message or sequence of messages to be communicated to the receiving terminal. The message may be of various types: (a) a sequence of letters as in a telegraph of teletype system; (b) a single function of time $f(t)$ as in radio or telephony etc.
2. A *transmitter* which operates on the message in some way to produce a signal suitable for transmission over the channel. In telephony this operation consists merely of changing sound pressure into a proportional electrical current. In telegraphy we have an encoding operation which produces a sequence of dots, dashes and spaces.
3. A *channel* which is merely the medium used to transmit the signal from transmitter to receiver. It may be wires, a coaxial cable, a band of radio frequencies, a beam of light, etc. Channels are imperfect and in the process transmitting signals, certain things are added to the signals which were not intended by the information source. These unwanted additions may be distortions of sound (in telephony, for example) or static (in radio), or distortions in shape or shading of picture (television), or errors in transmission (telegraphy), etc. All of these changes in the transmitted signal are called noise.
4. The *receiver* ordinarily performs the inverse operation of that done by the transmitter, reconstructing the message from the signal.
5. The *destination* is the person (or thing) for whom the message is intended (and, of course, the goal is that the received message is understood).

This model of communication made it possible to identify where in the process communication may fail. It allowed one to show unequivocally that a problem with encoding, channel, noise, and decoding can result in the failure of effective information transmission and retrieval. Thus, Shannon wanted to minimize the average number of bits needed to encode characters in messages that were sent through noisy channels to avoid miscommunication.

Let us discuss the problem of encoding a message, say English text, into binary digits in the most efficient way. The essential step is to assign probabilities to each of the conceivable messages in a way which incorporates the prior knowledge we have about the structure of English. So Shannon became the first person to consider communication as a **statistical process**. Nevertheless, in all likelihood we shall never know the 'true' probabilities of English messages; and so Shannon suggested the principle by which we may construct the probability distribution actually used for applications. His recipe was to choose to use some of our statistical knowledge of English in constructing a code, but not all of it. In such a case we consider the source with the maximum information subject to the statistical conditions we wish to retain. The information of this source determines the channel capacity which is necessary and sufficient.

Suppose the message is a linear string of symbols that is L characters long, where each character is drawn independently from an n-letter alphabet, with probability p_i. Let l_i $(i = 1, 2, \ldots, n)$ represent the number of times that the ith type of character is observed in the message. When L is large, the most likely L-letter message will have the composition $l_i = Lp_i$, and this occurs with probability

$$P = p_1^{l_1} \cdots p_n^{l_n} = p_1^{Lp_1} \cdots p_n^{Lp_n}. \tag{1.4}$$

In the limiting case of an alphabet having only a single letter, $P = 1$ and thus the probability of guessing the message is 100%. The larger the alphabet is, the smaller the value of P and the more unlikely we guess the message (and the greater the uncertainty of receiving a particular message). A goal of Shannon was to construct a measure for the amount of information associated with a message. His idea was that the amount of information gained from the reception of a message depends on how *likely* it is; the less likely a message is, the more information is gained upon its reception. Another way to see it is that information in communication theory relates not so much to what you do say, as to what you could say. That is, **information is a measure of one's freedom of choice when one selects a message**.

Then, any monotonic function of $1/P$ can be regarded as a measure of the information gained from the reception of the message. As was pointed out by Hartley the most natural choice is the logarithmic function. Shannon (1948) provided in turn additional reasons about the convenience of the logarithmic measure:

I. By its practical usefulness.
   Parameters of engineering importance such as time, bandwidth, number of relays, etc, tend to vary linearly with the logarithm of the number of possibilities. For example, adding one relay to a group doubles the number of possible states of the relays. It adds 1 to the base 2 logarithm of this number.

II. By the extensiveness property of information.
   We choose a logarithmic law in order to insure additivity of the information contained in independent situations. For example, two identical channels must be twice the capacity of one for transmitting information. Or, suppose we take the log base $b = 2$, in such a way a device with two stable positions (such as a relay or a flip-flop circuit) can store one bit of information. Thus, N such devices can store N bits, since the total number of possible states is 2^N and $\log_2 2^N = N$.

III. By its mathematical suitability.
   Many of the mathematical limit operations are simple in terms of the logarithm but would require clumsy restatement in terms of the number of possibilities.

In fact, Shannon (1948) proved that the only mathematical function that unifies the two intuitions about extensiveness and probability is the one that defines the information in terms of the logarithm of the inverse probability $1/P$ (see box 1.2):

$$H \equiv \log_b(1/P), \tag{1.5}$$

Box 1.2. Shannon's information entropy theorem

Shannon (1948) stated that it is reasonable to require of a measure $H(p_1, p_2, \ldots, p_{n\cdot})$ of how much 'choice' is involved in the selection of the event or of how uncertain we are of the outcome the following properties:

1. H should be continuous in the *pi*.
2. If all the p_i are equal, $p_i = 1/n$, then H should be a monotonic increasing function of n. With equally likely events there is more choice, or uncertainty, when there are more possible events.
3. If a choice is broken down into two successive choices, the original H should be the weighted sum of the individual values of H.

Theorem: The only H satisfying the three above assumptions is of the form:

$$H = -K\sum_{i=1}^{n} p_i \log_b p_i .$$

where K is a positive constant.

where b is the base of the logarithm used. Common values of b are 2, Euler's number e, and 10, and the corresponding units of entropy are the ***bits*** for $b = 2$, ***nats*** for $b = e$, and ***bans*** for $b = 10$. In turn, using equation (1.4), H becomes:

$$H = -L\sum_{i=1}^{n} p_i \log_b p_i . \tag{1.6}$$

The mathematical form of H given above is the same as the entropy function $\mathfrak{S}$ $(\{p_i\})$ of Gibbs and Boltzmann (that we will discuss in the next chapter), and so Shannon called it ***information entropy***. He interpreted H, in a general case in which we have a set of possible events whose probabilities of occurrence are $p_1, p_2, \ldots, p_n$, as a measure of how much 'choice' is involved in the selection of the event or of how uncertain we are of the outcome. Hence, H is also called ***uncertainty***.

To illustrate information entropy as a measure of choice or uncertainty suppose messages which are linear strings of length $L = 8$ chosen from the $n = 4$ alphabet $\{1,2,3,4\}$, i.e., characters are chosen among these four numbers. For example, let us consider these three messages:

(A) 1, 1, 1, 1, 1, 1, 1, 1
(B) 1, 1, 1, 1, 2, 2, 3, 4
(C) 1, 1, 2, 2, 3, 3, 4, 4

How do we order these three messages in terms of their information entropy H? After a short inspection we can say by intuition that message (A) has low information entropy, message (B) has medium information entropy and message (C) has high information entropy. But there is another question that is actually equivalent. Suppose a character is picked at random; how do you order these messages in terms of how easy it is for you to guess the number? So here is where knowledge comes into play. For sequence (A) you always are going to hit the number 1, so that's easy. For sequence (B) your best guess is to bet that it is a 1, let's say that it is of medium difficulty. Finally, in sequence (C) all the four numbers are equally likely and so it is hard to guess the number. Indeed, seemingly disordered sequences of symbols are much more unpredictable and thus potential carriers of information. We will not learn much new when capable of predicting the content of a message analyzing only a part from it. More quantitatively, we can compute the probability P guessing the message, using equation (1.4), and then obtaining H, using equation (1.5). Thus, for these sequences we have:

(A) $p_1 = 1, p_2 = 0, p_3 = 0, p_4 = 0$, then $P = 1^8 \cdot 0^\circ \cdot 0^\circ \cdot 0^\circ = 1 \cdot 1 \cdot 1 \cdot 1 = 1$ and thus $H = \log_2 1 = 0$;

(B) $p_1 = \frac{1}{2}, p_2 = \frac{1}{4}, p_3 = \frac{1}{8}, p_4 = \frac{1}{8}$, then $P = \left(\frac{1}{2}\right)^4 \cdot \left(\frac{1}{4}\right)^2 \cdot \left(\frac{1}{8}\right) \cdot \left(\frac{1}{8}\right) = \frac{1}{2^{14}}$ and thus $H = \log_2 2^{14} = 14$;

(C) $p_1 = \frac{1}{4}, p_2 = \frac{1}{4}, p_3 = \frac{1}{4}, p_4 = \frac{1}{4}$, then $P = \left(\frac{1}{4}\right)^2 \cdot \left(\frac{1}{4}\right)^2 \cdot \left(\frac{1}{4}\right)^2 \cdot \left(\frac{1}{4}\right)^2 = \frac{1}{2^{16}}$ and thus $H = \log_2 2^{16} = 16$.

Table 1.1 summarizes the properties of each of the above three messages.

The ***entropy rate*** r of a data source means the average number of bits per symbol needed to encode it, i.e., $r = H/L$. For a given language, the maximum number of bits that can be coded in each character, assuming each character of the alphabet sequence is equally likely, is called its ***absolute rate*** R. As we have just seen, for $n = 4$ we get $R = 2 = \log_2 4$ bits/character while for $n = 8$ we get $R = 3 = \log_2 8$ bits/character. Indeed, from equation (3.17) we see that $R = \max(H)/L = -n \cdot 1/n \cdot \log_2 (1/n) = \log_2 n$. For English, with a 26 letter alphabet (27 if we include the space between words), the absolute rate is $\log_2 27$, or about 4.75 bits/letter. However, it is known that the frequency of occurrence of the letters in English texts is not uniform but the one in table 1.2. Then, using Shannon's formula, equation (1.6), we can compute the entropy and we get (last row of table 1.2) $H = 4.11$. It turns out that the result is smaller than the absolute rate $R = 4.75$ but perhaps not as small as one would expect. Indeed, it is more convenient to work with a **normalized entropy** a.k.a. ***efficiency*** η given by:

$$\eta = \frac{H}{\max(H)} = \frac{H}{\log_b n}, \tag{1.7}$$

Table 1.1. Difficulty, probability of guessing P, information entropy H and rate $r = H/L$ for messages (A) to (C).

Message	Difficulty	P	H (in bits)	$r = H/L$ (in bits/character)
1, 1, 1, 1, 1, 1, 1, 1	Easy	1	0	0
1, 1, 1, 1, 2, 2, 3, 4	Medium	$1/2^{14}$	14	14/8 = 1.75
1, 1, 2, 2, 3, 3, 4, 4	Hard	$1/2^{16}$	16	16/8 = 2

Table 1.2. The frequencies of the letters in the English language.

i	Letter	p_i	$-\log_2 p_i$	$-p_i \log_2 p_i$
1	a	0.0575	4.1203	0.2369
2	b	0.0128	6.2877	0.0805
3	c	0.0263	5.2488	0.1380
4	d	0.0285	5.1329	0.1463
5	e	0.0913	3.4532	0.3153
6	f	0.0173	5.8531	0.1013
7	g	0.0133	6.2324	0.0829
8	h	0.0313	4.9977	0.1564
9	i	0.0599	4.0613	0.2433
10	j	0.0006	10.7027	0.0064
11	k	0.0084	6.8954	0.0579
12	l	0.0335	4.8997	0.1641
13	m	0.0235	5.4112	0.1272
14	n	0.0596	4.0685	0.2425
15	o	0.0689	3.8594	0.2659
16	p	0.0192	5.7027	0.1095
17	q	0.0008	10.2877	0.0082
18	r	0.0508	4.2990	0.2184
19	s	0.0567	4.1405	0.2348
20	t	0.0706	3.8242	0.2700
21	u	0.0334	4.9040	0.1638
22	v	0.0069	7.1792	0.0495
23	w	0.0119	6.3929	0.0761
24	x	0.0073	7.0979	0.0518
25	y	0.0164	5.9302	0.0973
26	z	0.0007	10.4804	0.0073
27	—	0.1928	2.3748	0.4579
Sum				$\mathbf{H = 4.1094}$

which measures the entropy relative to the 'optimized alphabet' whose symbols had uniform distribution. Hence, for English, $\eta = 4.11/4.75 = 0.864$.

In fact English text, treated as a string of characters, has fairly low entropy, i.e., is fairly predictable and its efficiency is smaller than 0.864. For example, let us consider the sentence 'The quick brown fox jumps over the lazy dog', which is an English-language pangram—a sentence that contains all of the letters of the English alphabet. The phrase is commonly used for touch-typing practice, testing typewriters and computer keyboards, displaying examples of fonts, and other applications involving text where the use of all letters in the alphabet is desired. Now, if we write:

Th qck brn fx jmpd ovr th lz dg

it can be easily decoded and recognized as the previous sentence but requiring only 75% of the letters in the complete text. It turns out that vowels are often unnecessary to the meaning of text messages. In general, vowels supply redundant information in the message and often may be discarded without loss of meaning.

Likewise, if we do not know exactly what is going to come next, we can be fairly certain that, for example, 'e' will be far more common than 'z'. The same happens for combinations of letters. For example, the combination 'qu' will be much more common than any other combination with a 'q' in it, etc. So ***redundancy*** $\mathcal{R}$ can be formally defined as

$$\mathcal{R} = 1 - \eta. \tag{1.8}$$

Indeed, the typical entropy rate of English text is 1.3 bits per letter of the message (Scheiner 1996), compared with an absolute rate of 4.75 bit/letter, meaning a redundancy of 3.45 bits/letter, i.e., for written English $\eta = 0.274$ and $\mathcal{R} = 0.726$.

1.3 Entropy as missing information

Shannon's entropy is often equated with information, but this is not correct; entropy indeed is missing information. As we have seen in the preceding section, recognizing that noise is inherent in real world channels, and thus doubt can arise in communication and lead to miscommunication, Shannon developed the notion of information entropy as a means of measuring such uncertainty and to statistically quantify how much information is contained within a received message. Indeed, originally, Shannon referred to the quantity he was seeking to define as 'choice,' 'uncertainty,' 'information' and 'entropy.' So, in this subsection, we want to make it clear about entropy as missing information.

The following example of guessing the location of a coin from *A Farewell to Entropy* by Arieh Ben-Naim (2008) about why the term 'choice' may be misleading and the term '***missing information***' (MI) is preferable is illuminating.

Suppose we are given n boxes and we are told that a coin was hidden in one box, and we have to guess in which box the coin is placed. We are also told that the box in which the coin was placed was chosen at random, i.e., with probability $1/n$.

The term 'choice' is easily understood in the sense that in this particular game, we have to *choose* between n boxes to place the coin. Clearly, for $n = 1$, there is only one box to choose and the amount of 'choice' we have is zero; the coin is in that box. It is also clear that as n increases, the larger n is, the larger the 'choice' we have to select the box in which the coin is to be placed. However, the interpretation of H as the amount of 'choice', for the case of unequal probabilities is less straightforward. For instance, if the probabilities of, say, ten boxes are 9/10, 1/10, 0, …, 0, it is clear that we have less choice than in the case of a uniform distribution, but in the general case of unequal probabilities, the 'choice' interpretation is not entirely satisfactory. The term 'information' is intuitively more appealing. If we are asked to find out where the coin is hidden, it is clear, even to the lay person, that we *lack information* on 'where the coin is hidden.' It is also clear that if $n = 1$, we need no information, we know that the coin is in that box. As n increases, so does the amount of the information we lack, or the MI. This interpretation can be easily extended to the case of unequal probabilities. Clearly, any non-uniformity in the distribution only increases our information, or decreases the MI. In fact, all the properties of H listed in box 1.2 are consistent with the interpretation of H as the amount of MI.

With the computation of the MI in mind, we can think of the information content of a message per character as the number of yes/no questions we would have to ask to guess what character was picked. So imagine a guessing game, where someone picks one of these numbers and the questioner has to guess the number. Asking those questions is equivalent to pinpoint or decoding the message received. A naive approach for the above alphabet {1,2,3,4} would be to ask four questions: Question 1: Is it 1? If not, question 2: Is it 2? If not, question 3: Is it 3? If not, question 4: Is it 4? As the reader has noticed, the fourth question is unnecessary since if the number is different from 1, 2 and 3, the only chance it has is to be equal to 4. Therefore, it seems we would need three questions. Actually we can do even better: a general procedure to find the minimal average number of questions per character necessary to guess the character is to use a binary search tree. The binary tree for the alphabet {1,2,3,4} looks like figure 1.1(a). We see that such a binary tree would reveal the label of the digit using a minimum number of questions, that is, just two. If instead of an alphabet of four characters we have one of eight characters, i.e., a message can be any combination of the numbers {1, 2, 3, …, 8} and all eight numbers are equally probable, the binary search tree looks like figure 1.1(b). Likewise, this binary search tree would reveal the label of the digit using three questions.

Let us apply the binary search tree procedure we have seen in the previous section to a couple of examples to acquire the total MI. That is, given a system which can be in several possible configurations or ***microscopic states*** or ***microstates***, by asking questions we acquire information on this configuration. The larger the MI, the larger the average number of questions to be asked. The MI is the minimum number of binary (yes or no) questions we need to ask to know the specific microscopic configuration of the system. Both examples, besides serving to illustrate how to acquire and quantify MI, will be illuminating for another important issue. Namely, for a system such that its ***macroscopic state*** or ***macrostate*** is specified, in terms of

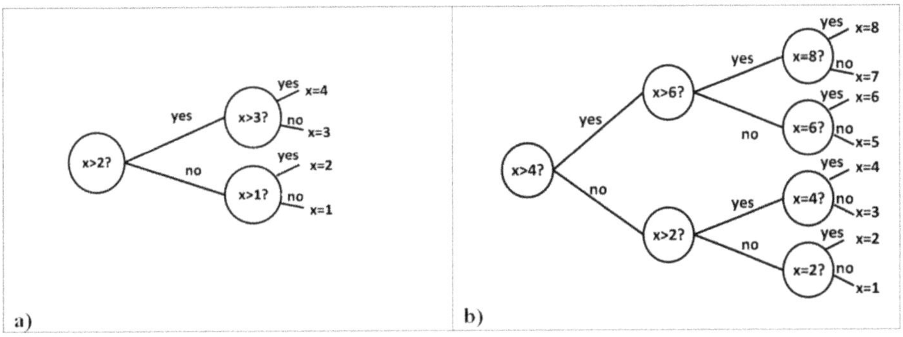

Figure 1.1. Binary search trees. (a) For an alphabet of $n = 4$ characters. (b) For an alphabet of $n = 8$ characters.

global or average quantities, which is the MI we need if we want to guess its microscopic state (a central theme of the next chapter).

Example 1: MI of two fair dice with even sum.

Suppose that we are told that in throwing two fair dice the sum of the outcomes is an even number. What is the missing information (MI)?

Thus, in this example the macroscopic state is the sum of the outcomes of the pair of dice and the microscopic state is the specific pair of values for each die. There are altogether 36 possible outcomes, 18 of which have a sum that is an even number. Therefore, the probability of each of these 18 states whose sum is even is $p_i = 1/18$, and thus $H = -18 \times (1/18) \times \log_2(1/1/18) = \log_2 18 = 4.17$.

Let us compute the MI and show that it is equal to H. Using a binary search tree we obtain that the average minimum number of questions is given by:

$$\frac{4}{9} \times 4 + \frac{5}{9} \times \left[2 + \frac{2}{5} \times 2 + \frac{3}{5} \times \left(1 + \frac{1}{2} + \frac{1}{2} \times 2\right)\right] = \frac{375}{90} = 4.17.$$

Therefore, we conclude that MI $= H = 4.17$ bits.

Had we no information about the sum of the two fair dice, there would be 36 possible combinations for the outcome. Since all the 36 results are equi-probable, i.e., $p_i^0 = 1/36$ for all I (we denote by p_i^0 the probability with zero information on the outcome), it turns out that $H^\circ = -36 \times (1/36) \times \log_2(1/1/36) = \log_2(36) = 5.17$ bits. And, again, a similar computation of the MI using a binary search tree gives MI° $= H^\circ = 5.17$ bits.

Notice that MI° – MI $= H^0 - H = \log_2 36 - \log_2 18 = \log_2(36/18)$. That is, knowing that the sum is even we have gained $\log_2(36/18) = \log_2 2 = 1$ bit.

So, a general lesson of this example is that in cases in which there are initially N equally probable choices, and we are told something that narrows the possibilities down to one of M equally probable choices, then the information we gain is $\log(1/(M/N)) = \log(N/M)$ bits.

We can therefore state the following rule:

If we receive information that narrows down a set of N equi-probable choices to one of M equi-probable choices, then we have gained $\log(N/M)$ bits of information.

Example 2: Microstates, macrostates and MI of a set of balls in boxes.

The first example we consider, besides serving to illustrate how to acquire and quantify MI, will be useful for connecting the issue of MI for a system such that its macroscopic state is specified and we want to guess its microscopic state (a central theme of the next chapter). Thus, imagine a game involving a system of N boxes; for simplicity we take $N = 2^k$, where k is an integer. Each box can be either empty or occupied by a single ball, corresponding to the state denoted by 0 or 1 respectively. The rule of the game is that one box is taken randomly and its corresponding state is inverted: if the box is empty a ball is put into the box and in the opposite case the ball is removed from the box (i.e. from 0 to 1 or vice versa). The macroscopic state or configuration corresponds here to the sum n of the binary numbers over all the N

boxes (equal to the number of balls). Thus, there are $N + 1$ macroscopic states: $n = 0, 1, \ldots, N$. An arbitrary macroscopic configuration, characterized by a sum equal to n, corresponds to:

$$W(n) = \binom{N}{n} = \frac{N!}{n!(N-n)!} \tag{1.9}$$

microscopic configurations, since $W(n)$ is the number of ways of placing n identical objects in N distinguishable boxes (with no more than one ball per box).

If we start with all boxes empty (all 0s), the sum is initially 0 and there is 100% chance that the sum increases at the next turn. Then, at the next turn, because there are still many more 0's than 1's, there is still a higher probability to change a 0 to 1 and to increase the sum than the inverse. Then, with the sum increasing the probability of increasing the sum will decrease but will still be higher until the sum reaches $N/2$. In that case, there are as many empty as filled boxes and 50% chance to increase or decrease the sum at the next turn. After that, the sum will fluctuate around $N/2$ but will almost never deviate significantly; there is a very low probability for the sum to go back to 0. We could have started from all boxes filled (all 1s) and the sum equals N but the result would have been the same: the sum would slowly equilibrate toward $N/2$. This is because there is only one microscopic configuration for the macroscopic configuration of 0-sum (or N-sum), but many specific configurations for the macroscopic configuration of $N/2$-sum. The random evolution implies that the system will spend more time in the most populated macroscopic configuration ($N/2$-sum) than in the starting configuration.

MI is the minimum number of binary questions we need to ask, knowing the macroscopic configuration, to know the specific microscopic configuration. In the starting configuration of all 0s, we know that the sum is zero and we do not need to ask any question to know the specific configuration: the MI is zero. At the next turn, the sum is 1 and there are N possible microscopic configurations. To know exactly in which box the ball is we proceed in the same way as we did to guess the sequence of numbers of the message considered in the previous section: we use the binary search tree. We suppose the N boxes arranged in a row and we start asking 'Is the ball in the right half of the row?' This allows us to discard $N/2$ boxes. Our next question is 'Is the ball in the right half of the remaining $N/2$?' And so on and so forth until we end with just one box. Thus we need to ask a minimum of $\log_2 N$ binary questions of the kind, and this is therefore the MI.

Let us go one step further and compute the MI when there are two balls hidden in N boxes. Again, we know that the boxes were chosen at random. For the first ball, we have one out of N boxes; for the second, we have one out of $(N - 1)$ boxes, etc. Therefore, according to equation (1.9), we obtain $W(2) = N!/[2!(N-2)!] = N(N-1)/2$. A calculation of the MI is much cumbersome than for one ball but it can be shown that the average number of questions coincides with $\log_2 W(2) = \log_2[N(N-1)/2] = \log_2[2^{k-1}(2^k - 1)] = k{-}1 + \log_2(2^k - 1)$. And the MI of a general macroscopic configuration, characterized by a sum n, by equation (1.9) is always $\log_2 W(n)$. Thus, the higher W, the higher the MI.

In summary:

- **Information entropy H measures the amount of missing information (MI) for a complete probabilistic description of a system.**
- **As we have seen, H = MI corresponds to the mean number of binary questions that must be asked in an optimal decision strategy to determine the state of a particular realization given the state of the ensemble to which it belongs.**

1.4 Working with incomplete information: the principle of maximum entropy to find minimally prejudiced distributions

Imagine we want to explain or predict the behavior of a real system when its conditions are changed. For example, let us consider a gas in a cylinder with a piston. As it happens with any real system, the knowledge or information we have about the state is in general incomplete. For example, we can know a series of aggregate or average properties, like the pressure P, volume V and temperature T of the gas—i.e., the thermodynamic state or macrostate $\mathcal{S}$ (P,V,T). Nevertheless, from the point of view of Newton's classical mechanics a true description of the gas would imply knowing the position $\mathbf{x}_i$ and momentum $\mathbf{p}_i = m_i\mathbf{v}_i$ (m_i is the mass of molecule i and $\mathbf{v}_i$ its velocity) of each molecule i of the gas, or its microscopic state $s(\mathbf{x}_1,\mathbf{x}_2,\ldots,\mathbf{x}_N,\mathbf{p}_1, \mathbf{p}_2,\ldots,\mathbf{p}_N)$. In normal conditions of P, V and T one has of the order of $N = 10^{22}$ molecules and thus such complete microscopic description is definitely out of reach (besides not being very useful).

A microscopic treatment is still possible provided we consider probabilities for each microscopic state s; this is the subject of *statistical mechanics* (treated in chapter 2). Statistical mechanics is a mathematical framework that applies statistical methods and probability theory to large assemblies of microscopic entities (e.g., molecules or atoms). It does not assume or postulate any natural laws, but explains the macroscopic behavior of Nature from the behavior of such ensembles. The primary goal of statistical mechanics is to derive the classical thermodynamics of materials in terms of the properties of their constituent particles and the interactions between them. In fact, statistical mechanics allows one to express the macroscopic quantities—like P, V and T—in terms of averages over the microscopic states $s(\mathbf{x}_1,\mathbf{x}_2,\ldots,\mathbf{x}_N, \mathbf{p}_1,\mathbf{p}_2,\ldots,\mathbf{p}_N)$ that describe the motions occurring inside the material. This kind of approach, providing a connection between the macroscopic properties of a system, and the microscopic behaviors of its elementary constituents has shown its usefulness beyond physical systems; in later chapters we will see applications in biology as well as in economics.

Thus, probabilities are introduced because of incomplete knowledge of the state of a system. Yet a common problem we have to deal with is that the probability distribution of microstates $\mathcal{P}$ (s) is not known, and there are many different probability distributions which are compatible with the available (incomplete) data on the system of interest.

The maximum entropy (MaxEnt) principle is a general method to make the least biased inferences compatible with available data, in the form of a set of known constraints proposed by E T Jaynes (1957a, 1957b) (figure 1.2).

Figure 1.2. Edwin Thompson Jaynes (1922–98). Photo courtesy of Larry Bretthorst of Washington University.

But, before discussing the MaxEnt method and its main implications, it is necessary to open a parenthesis to briefly present two main interpretations or uses of probability in the next subsection.

1.4.1 Frequentist or physical probability versus subjective or Bayesian probability

The word 'probability' has been used in a variety of ways and senses since it was first applied to the mathematical study of games of chance (Hacking 1975). There are two main interpretations for probability: the so-called 'frequentist' or 'physical' and the 'subjective' or Bayesian probabilities.

In the frequentist interpretation, probabilities are associated with physical set-ups such as gambling devices, like flipping coins or rolling dice, or natural physical systems, e.g., radioactive atoms. Such probabilities are estimated as the fraction of times an outcome is observed in a large number of random trials. Thus, the frequentist interpretation views probability as the limit of the relative frequency of an event after many trials (Hájek 2002). For instance, the frequency of appearance of each face of a die is readily determined by rolling a die many times, then dividing the number of appearances of each outcome by the total number of dice rolls. Large sequences of many repeated trials demonstrate that the empirical frequency converges to the limit 1/6 as the number of trials goes to infinity. Thus, a limitation of this frequentist view is that it is of course impossible to actually perform infinite repetitions of a random experiment to determine the probability of an event. For example, when we wrote the probability P of a message, equation (3.16), we assumed that we see a large number of messages, from which we can compute the frequencies $p_i = l_i/L$ of the different letters. But what if only one message is seen? Actually, Bernoulli realized long ago that the enumeration of options may be done in very few cases and almost nowhere else than in games of chance.

On the other hand, in the ***Bayesian*** or ***subjective interpretation*** of probabilities the concept of probability is not limited to situations that are replicable. Furthermore, probabilities can be assigned to any statement whatsoever, even when no random process is involved. In this subjective interpretation, probability expresses a degree of belief in an event of an observer's based on prior knowledge, such as the results of previous experiments, or on personal beliefs about the event (Cox 1961). That is, probability corresponds to a state of knowledge, and the rules of probability are simply ways to draw inferences from premises. For instance, the probability of rain tomorrow, say p_i, is a quantity that can be estimated, even though it is not describable by a repeatable experiment. In this instance it makes no sense to speak of a ratio such as $p_i = l_i/L$ because the number of times it rains tomorrow is not an enumerable quantity. The subjective view is evidently the broader one since it is always possible to interpret frequency ratios in this way.

Jaynes indeed adopted this more broadly ***subjective*** viewpoint as a description of a state of knowledge. He regarded MaxEnt as a way of minimizing the extent to which we are incorporating into our inferences assumptions that we have no basis for making.

The recipe of MaxEnt to devise the probability distribution $\mathcal{P}(s)$ among the many probability distributions which are compatible with the known information is to choose the one that maximizes the information entropy subject to the known constraints. Jaynes thus showed that this prescription allows deriving the statistical mechanics relations—that we present in the next chapter—in a very elementary way. Importantly, this derivation doesn't require additional assumptions not contained in the laws of mechanics (such as ergodicity, equal *a priori* probabilities, etc).

Perhaps most importantly, Jaynes also realized that although the information entropy of Shannon was intended for use in communication theory, it is a much more general concept related with our uncertainty or ignorance about the state of an arbitrary system (like a gas of molecules, an ecosystem or a market). Thus, information theory provides a constructive criterion for setting up probability distributions on the basis of partial knowledge, and leads to a type of statistical inference which is called the maximum-entropy estimate. It is the least biased estimate possible on the given information. In other words, **the maximization of entropy is a method of reasoning which ensures that no unconscious arbitrary assumptions have been introduced** (Jaynes 1957a).

This important insight has led to regarding MaxEnt in a much broader light, as a method which is applicable in many different areas beyond communications and physics; from problems in the fields of biology (see chapter 3), neuroscience, to finances (see chapter 4) and social sciences, such as sociology. That is, Jaynes discovered that the MaxEnt principle was powerful enough to be used as a statistical tool in its own respect, independent of thermodynamics by using information entropy.

1.4.2 MaxEnt as a method of making predictions from limited data by assuming maximal ignorance

The information entropy for an arbitrary system can be written by an expression similar to equation (1.6) (Shannon 1948, Jaynes 1957a):

$$H = -\mathcal{N}\sum_{s=1}^{S} P(s)\log\ P(s), \tag{1.10}$$

where $\mathcal{N}$ is a normalization constant, S is the number of sates of the system and $\mathcal{P}(s)$ denotes the probability of the state s of the system. In fact, by writing equation (1.10) we assumed a discrete set of states for the system, i.e., s is an integer number. In the case we have a continuum of states, described by a set of K continuum variables $\mathbf{v} = [v_1, v_2, \ldots, v_K]$ (as in the example of the gas presented at the beginning of this section), we can denote the corresponding multivariate probability distribution of K random variables by $\mathcal{P}(\mathbf{v})$, and thus instead of a sum H is expressed as an integral:

$$H = -\int \mathrm{d}\mathbf{v} P(\mathbf{v}) \log P(\mathbf{v}) = -\int \prod_{i=1}^{K} \mathrm{d}v_i P(v) \log \mathbf{P}(\mathbf{v}), \tag{1.10'}$$

The idea underneath MaxEnt is that in most real systems, the information entropy is not free to be at its maximum unconstrained value. Instead, it is checked by many constraints, For example, in the case of the gas we know it must obey the *law of conservation of energy*.

However, a common problem is that the information we have about the system is limited, and thus many different probability distributions are compatible with the data. Of all such distributions, the choice is made by finding the set $\{\mathcal{P}(s)\}$ that both maximizes the entropy and satisfies the known conditions. Such distribution is certainly the least biased distribution compatible with the data. Since with $\mathcal{P}(s)$ the quantities that we can build are expected values, we express the conditions in the form of constraints on expected values. Suppose for example that we know that certain quantity A takes the value $\bar{A}$:

$$\sum_{s=1}^{S} P(s) A(s) = \bar{A}, \tag{1.11}$$

or

$$\int P(\mathbf{v}) A(\mathbf{v}) \mathrm{d}\mathbf{v} = \int \prod_{i} \mathrm{d}v_i A(v) P(\mathbf{v}) = \bar{A}. \tag{1.11'}$$

For instance, a first constraint to be imposed always is the normalization condition

$$\sum_{s=1}^{S} P(s) = 1, \tag{1.12}$$

or

$$\int P(\mathbf{v}) \mathrm{d}\mathbf{v} = \int \prod_{i} \mathrm{d}v_i P(\mathbf{v}) = 1, \tag{1.12'}$$

i.e., $A(s) = 1 = \bar{A}$ for all states s.

A well-known analytical technique in mathematical optimization, the *method of Lagrange multipliers* (Arfken 1985), is a strategy for finding the local maxima and minima of a function subject to equality constraints (i.e., subject to the condition that one or more equations have to be satisfied exactly by the chosen values of the variables). It is named after the mathematician Joseph-Louis Lagrange. The basic idea is to convert a constrained problem into a form such that the derivative test of an unconstrained problem can still be applied. Thus, to obtain the maximum entropy probability distribution $\mathcal{P}(s)$ consistent with the constraint (1.12) and C additional different constraints of the general form (1.11) we apply the method of the Lagrange multipliers, which converts a constrained maximization problem into an unconstrained one by maximizing the new function H' given by:

$$H' = H - \lambda_0\left(\sum_{s=1}^{S} P(s) - 1\right) - \sum_{i=1}^{C}\lambda_i\left(\sum_{s=1}^{S} P(s)A_i(s) - \bar{A}_i\right), \tag{1.13}$$

considered as a function of $\mathcal{P}_s$ and the Lagrange multipliers denoted by the Greek letter lambda: λ_0 is a Lagrange multiplier that assures the normalization of the $\mathcal{P}_s$ and each λ_i is a Lagrange multiplier enforcing each known value of the average of a quantity A_i (with $i = 1, \ldots, C$, i.e., we have one λ_i multiplier for each additional known constraint). This is done straightforwardly by solving the equation

$$\delta\left[H - \lambda_0\left(\sum_{s=1}^{S} P(s) - 1\right) - \sum_{i=1}^{C}\lambda_i\left(\sum_{s=1}^{S} P(s)A_i(s) - \bar{A}_i\right)\right] = 0, \tag{1.14}$$

where the variation is with respect to each $\mathcal{P}(s)$. The solution can be written as:

$$\boxed{P(s) = \frac{\exp\left(-\sum_{i=1}^{C}\lambda_i A_i(s)\right)}{Z(\lambda_1, \ldots \lambda_C)},} \tag{1.15}$$

where the denominator is a normalizing constant $Z(\lambda_1,\ldots,\lambda_C)$, denoted as *partition function*, which is given by:

$$Z(\lambda_1, \ldots, \lambda_C) = \sum_{s=1} e^{-\sum_{i=1}^{C}\lambda_i A_i(s)}. \tag{1.16}$$

The Lagrange multiplier λ_0 is already considered because equations (1.15) and (1.16) imply the normalization constraint. The other C λ_i parameters are determined from the solution of the C nonlinear equations:

$$\overline{A_i} = -\frac{\partial \ln Z}{\partial \lambda_i}, \tag{1.17}$$

which is identical to substitute equations (1.15) and (1.16) into (1.12). These equations are solved either algebraically or numerically.

To illustrate MaxEnt at work in box 1.3 we consider a very simple system: a die with six sides numbered from 1 to 6. The only thing we know about this die is that after throwing it thousands of times the mean of outcomes is 3.5. What are the probabilities which we have to assign to each side of the die?

Box 1.3. The probability distribution associated to a die

We have a die with six sides numbered from 1 to 6. The only thing we know about this die is that after throwing it thousands of times the mean of outcomes is $\bar{k}$. What are the probabilities which we have to assign to each side of the die?

The first step to solve this problem is to notice that we have two constraints:

$$\sum_{k=1}^{6} p(k) = 1, \tag{i}$$

$$\sum_{k=1}^{6} kp(k) = \bar{k}, \tag{ii}$$

i.e., we have only two equations and six unknown quantities. Therefore, there are many (infinite) solutions to this problem. For example, two possible solutions are:

$$p^{(1)} = \left\{\frac{6-\bar{k}}{25}, \frac{6-\bar{k}}{25}, \frac{6-\bar{k}}{25}, \frac{6-\bar{k}}{25}, \frac{6-\bar{k}}{25}, \frac{\bar{k}-1}{5}\right\},$$

$$p^{(2)} = \left\{\frac{6-\bar{k}}{5}, 0,0,0,0, \frac{\bar{k}-1}{5}\right\},$$

since it can be easily checked that both verify the constraints.

According to equation (1.13), we define the auxiliary function

$$\begin{aligned} H' &= H(p_1, \ldots, p_6) - \lambda_0\left(\sum_{k=1}^{6} p(k) - 1\right) - \lambda_1\left(\sum_{k=1}^{6} kp(k) - \bar{k}\right) \\ &= -\sum_{k=1}^{6} p(k)\log p(k) - \lambda_0\left(\sum_{k=1}^{6} p(k) - 1\right) - \lambda_1\left(\sum_{k=1}^{6} kp(k) - \bar{k}\right), \end{aligned} \tag{iii}$$

Taking the partial derivatives of H' with respect to each of the $p(k)$ we have:

$$-\log p(k) - 1 - \lambda_0 - \lambda_1 k = 0,$$

which leads to

$$p(k) = \exp(-1 - \lambda_0 - \lambda_1 k) = \frac{\exp(-\lambda_1 k)}{Z(\lambda_1)}, \tag{iv}$$

with

$$Z(\lambda_1) = \sum_{k=1}^{6} e^{-\lambda_1 k} \tag{v}$$

The coefficient λ_1 in (iv) can be determined from the constraint:

$$\bar{k} = \sum_{k=1}^{6} kp(k) = \frac{\sum_{k=1}^{6} ke^{-\lambda_1 k}}{Z(\lambda_1)}. \tag{vi}$$

Thus, dividing (vi) by (v) we obtain:

$$\bar{k} = \frac{\sum_{k=1}^{6} ke^{-\lambda_1 k}}{\sum_{k=1}^{6} e^{-\lambda_1 k}} = \frac{\sum_{k=1}^{6} kx^k}{\sum_{k=1}^{6} x^k}, \tag{vii}$$

where $x = e^{-\lambda 1}$. That is, (vii) is equivalent to the algebraic equation of degree 6:

$$\sum_{k=1}^{6} (k - \bar{k})x^k = 0. \tag{viii}$$

Let us consider two particular cases:

(A) $\bar{k} = 3.5$. In this case the equation (viii) has only one real root: $x = 1$, i.e., $\lambda_1 = 0$. Therefore, by (iv) all p_k are equal, and the only solution we have is:

$$p(k) = 1/6 \text{ for all } k, \tag{ix-a}$$

as shown in figure 1.3(a).

Thus, the value of H is maximized when all the faces of the die have the same probability. Indeed this result could have been anticipated, since 3.5 = (1 + 2 + 3 + 4 + 5 + 6)/6 is the average corresponding to a fair die and this is consistent with the notion that the distribution is uniform, but it is also consistent with several other possibilities. For example, the possibility that $p(1) = p(2) = p(5) = p(6) = 0$ and $p(3) = p(4) = 1/2$. Which should we assume? Logically, it makes no sense to assume a loaded die for which only 3s and 4s will appear when we have no reason to suspect that is the case.

(B) How things change if $\bar{k} = 4.5$? Now, the only real solution is $x = 1.449\,25$.

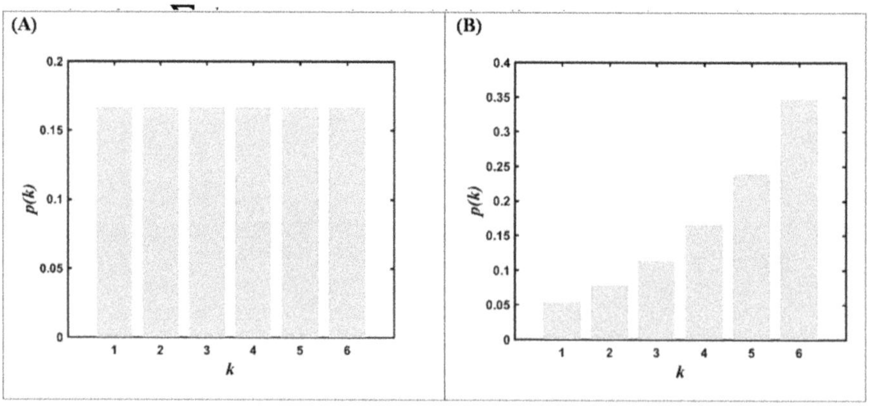

Figure 1.3. The maximum probability distribution for a die. (a) Mean $\bar{k} = 3.5$, which is consistent with a fair die. (b) Mean $\bar{k} = 4.5$, implying a loaded die.

which in turn implies that the probability distribution, shown in figure 1.3(b), is given by:

$$p(k) = 0.0375 \times 1.449\,25^{k}. \tag{ix-b}$$

That is, the die is loaded, and the likelihood of each face grows exponentially with its value.

To conclude let us see other examples of probabilities distributions of a random variable X produced by MaxEnt assuming increasing number of constraints.

- First, consider the case of no constraints, i.e., $\lambda_i = 0$ for all i. Then, according to equation (1.15) we have

$$P(x) = \text{constant} = \frac{1}{Z(\lambda_0)} = e^{-1-\lambda_0},$$

If the values that the variable X can take are discrete, to satisfy our the constraint that all probabilities sum to 1, i.e., equation (1.12), then

$$P(x) \equiv P(s) = \frac{1}{S}. \tag{1.18a}$$

- Second, let us say that we know the mean of a continuous variable varying over a region R: $\int_R \mathrm{d}x\, x\mathbf{P}(x) = \mu$, i.e., we have one constraint $A_1(x) \equiv x$. Then, according to equation (1.15) we have

$$P(x) = \frac{\exp(-\lambda_1 x)}{Z(\lambda_1)}..$$

And, after solving for the Lagrange multipliers we obtain:

$$P(x) = \frac{\mathrm{e}^{-(x/\mu)}}{\mu}. \tag{1.18b}$$

- Third, if we know the mean and variance of $\mathcal{P}(x)$. In other words we have two constraints $A_1(x) \equiv x$ and $A_2(x) \equiv (x - \mu)^2$. Then, according to equation (1.15) we have

$$P(x) = \frac{\exp(-\lambda_1 x - \lambda_2 (x - \mu)^2)}{Z(\lambda_1, \lambda_2)}.$$

Thus a Gaussian has appeared. The condition that $\int_R \mathrm{d}x\mathbf{P}(x)$ be finite requires $\lambda_1 = 0$ and $\lambda_2 < 0$. After solving for the Lagrange multipliers we obtain the Normal distribution:

$$P(x) = \frac{\mathrm{e}^{-((x-\mu)^2/2\sigma^2)}}{\sqrt{2\pi\sigma}} \tag{1.18c}$$

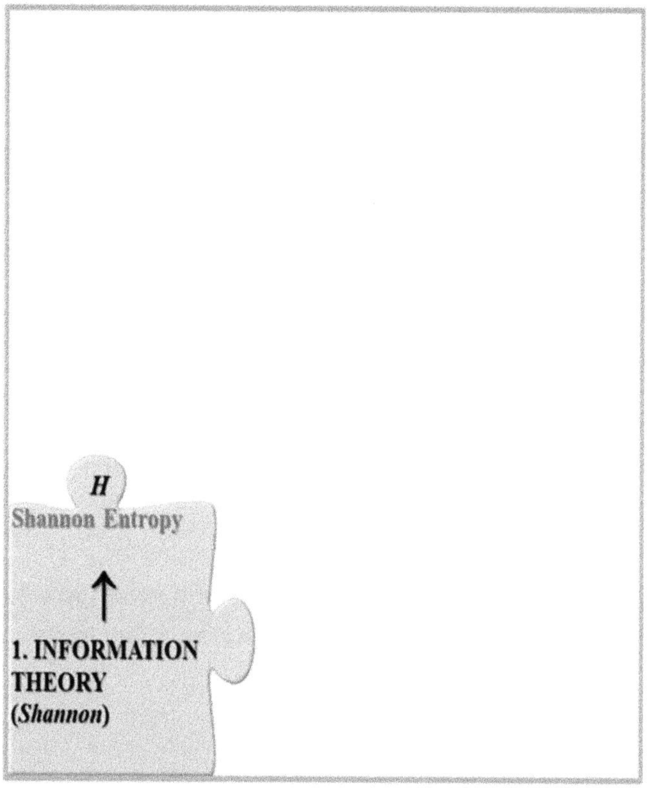

References

Adriaans P 2020 Information *The Stanford Encyclopedia of Philosophy (Fall 2020 Edition)* ed E N Zalta (Stanford,CA: Stanford University) https://plato.stanford.edu/archives/fall2020/entries/information/ The citation above refers to the version in the following archive edition: Fall 2020 (substantive content change)

Alon U 2006 *An Introduction to Systems Biology: Design Principles of Biological Circuits* (London/Boca Raton, FL: Chapman and Hall/CRC Press)

Arkfen G 1985 *Mathematical Methods for Physicists* 3rd edn (New York: Academic)

Ben-Naim A 2008 *A Farewell to Entropy: Statistical Thermodynamics Based on Information* (Singapore: World Scientific)

Brillouin L 1962 *Science and Information Theory* 2nd edn (New York: Academic)

Camazine D, Franks S and Theraulaz B 2003 *Self-Organization in Biological Systems* (Princeton, NJ: Princeton University Press)
Castells M 1996 *The Information Age: Economy, Society and Culture* (Oxford: Blackwell)
Capurro R and Hjørland B 2003 The concept of information *Ann. Rev. Inform. Sci. Technol. (ARIST)* **37** 343–411
Chiu E, Lin J, Mcferron B, Petigara N and Seshasai S 2001 Mathematical theory of Claude Shannon. A study of the style and context of his work up to the genesis of information theory *Commun. Bell Syst. Tech. J.* **27** 379–423
Cox R T 1961 *The Algebra of Probable Inference* (Baltimore, MD: John Hopkins University Press)
Crick F H 1958 On protein synthesis *Symposia of the Society for Experimental Biology, Number XII: The Biological Replication of Macromolecules* ed F K Sanders (Cambridge: Cambridge University Press) pp 138–63
Davies P 2020 Does new physics lurk inside living matter? *Phys. Today* **73** 34–40
Golan A 2006 *Information and Entropy Econometrics—A Review and Synthesis* (Hanover, MA: now Publishers)
Hacking I 1975 *The Emergence of Probability* (Cambridge: Cambridge University Press)
Hájek A 2002 Interpretations of probability *The Stanford Encyclopedia of Philosophy* ed E N Zalta (Stanford, CA: Stanford University)
Hartley R V L 1928 Transmission of information *Bell Syst. Tech. J.* **7** 535–63
Hayek F 1937 Economics and knowledge *Economica* **IV** 33–54
Hayek F 1945 The use of knowledge in society *Am. Econ. Rev.* **35** 519–30
Hurwicz L 1973 The design of mechanisms for resource allocation *Am. Econ. Rev.* **63** 1–30
Jaynes E T 1957a Information theory and statistical mechanics I *Phys. Rev* **106** 620–30
Jaynes E T 1957b Information theory and statistical mechanics II *Phys. Rev* **108** 171–90
Jutterström M 2018 *Markets as Open Systems: Organizing and Reorganizing a Financial Market* Oxford Scholarship Online. Available at: https://oxford.universitypressscholarship.com/view/10.1093/oso/9780198815761.001.0001/oso-9780198815761-chapter-8
Leavitt S A 2004 *Deciphering the Genetic Code: Marshall Nirenberg. The Coding Craze* (Stetten Museum, Office of NIH History) Retrieved 2021-26-03 from: https://history.nih.gov/display/history/Nirenberg+Introduction
Lewis G N 1930 The symmetry of time in physics *Science* **71** 569–77
Maxwell J C 1871 *Theory of Heat* (London and New York: Longmans) Reprinted (2001) New York: Dover
Nicolis G and Prigogine I 1977 *Self-organization in Nonequilibrium Systems: From Dissipative Structures to Order Through Fluctuations* (New York: Wiley)
Nyquist H 1924 Certain factors affecting telegraph speed *Bell Syst. Tech. J.* **3** 324–46
Scheiner B 1996 *Applied Cryptography* (Hoboken, NJ: John Wiley & Sons, Inc.)
Shannon C E 1948 A mathematical theory of communication *Bell Syst. Tech. J.* **27** 379–423
Shannon C E and Weaver W 1949 *The Mathematical Theory of Communication* (Champaign, IL: University of Illinois Press)
Voit J 2003 *The Statistical Mechanics of Financial Markets* 2nd edn (New York: Springer)
Zaman M (ed) 2009 *Statistical Mechanics of Cellular Systems and Processes* (Cambridge: Cambridge University Press)

Chapter 2

The synthesis of information theory and thermodynamics: Shannon entropy and Boltzmann entropy are the same thing

'If we are describing only a state of knowledge about a single system, then clearly there can be nothing physically real about frequencies in the ensemble; and it makes no sense to ask, 'which ensemble is the correct one?'... Gibbs understood this clearly; and that, I suggest, is the reason why he does not say a word about ergodic theorems,...'

— E T Jaynes (1978)

'...there comes a time when for every addition of knowledge you forget something that you knew before. It is of the highest importance, therefore, not to have useless facts elbowing out the useful ones.'

— Arthur Conan Doyle, *A Study in Scarlet*

We start by reviewing statistical mechanics. Statistical mechanics originated in the mid-19th century in an effort to explain the phenomenological laws of thermodynamics from the more fundamental Newton's laws mainly with the works of Maxwell, Boltzmann and Gibbs.

Next, we show how the statistical mechanics of Boltzmann and Gibbs can be derived as a straightforward application of using the general maximum entropy (MaxEnt) formalism. This provides a connection between Shannon's information entropy and Boltzmann's thermodynamic entropy which has been instrumental for extending statistical physics to complex systems beyond the realm of physics, from stock markets, to societies or ecosystems.

doi:10.1088/978-0-7503-3931-5ch2

We conclude the chapter with a paradox posed by Maxwell more than 150 years ago, the so-called Maxwell's demon paradox, based on a thought experiment. The prevailing wisdom, until some decades ago, was that the transfer of information from an object to be measured (in the case of Maxwell's demon paradox, a molecule) to a meter or register requires energy dissipation. However, thanks to the works of Landauer and Bennett, it was later realized that the dissipation required to save the second law and to prevent us from making molecules in thermal equilibrium do work comes not from the measurement process but has a different origin connected with information storing, more precisely with the erasure of information. This paradox is important because it was contemplation of Maxwell's demon that clarified the link between information and entropy as well as realizing that information can be transformed into energy and thus treated as a physical quantity.

2.1 Basics of statistical physics

Statistical mechanics originated in the mid-19th century in an effort to explain the phenomenological laws of thermodynamics from the more fundamental Newton's laws (and this is why it is also frequently referred to as ***statistical thermodynamics***). It has been used since then by physicists to analyze systems too complex to tackle by the canonical approach of solving the corresponding deterministic equations for their constituents. Indeed, an entire new branch of physics, ***statistical physics***, evolved from statistical mechanics. Statistical physics deals with large populations through methods of probability theory and statistics to solve a wide variety of physical problems.

Later, statistical physics has proved to be a powerful method to analyze complex systems beyond the realm of physics. For example, it has proven to be fruitful for the investigation of what is nowadays called 'soft matter', which encompasses complex objects like colloids, membranes and biomolecules. Statistical physics has also found applications in other more distant disciplines such as ecology (Maynard Smith 1978, Harte 2011, Fort 2013), economics (Lux 2007), finances (Ausloos *et al* 1999, Voit 2005) and social sciences (Castellano *et al* 2009). MaxEnt has been instrumental for this expansion outside physics.

The connection between information entropy and thermodynamic entropy, provided by the Maximum Entropy (MaxEnt) method of inference from information theory, has been instrumental for extending statistical physics to these fields.

2.1.1 The program of statistical physics: from microphysics to macrophysics

The essence of the statistical physics program is to explain a 'macroscopic' description of a system, involving aggregate variables and known phenomenological rules connecting them, in terms of the 'microscopic' components that constitute the system and the fundamental laws they obey. Indeed, the original goal was to explain the gas laws. The gas laws were crucial for understanding how to convert heat to work during the industrial revolution. As an illustrative example, imagine a mol of gas in a container of volume V, at pressure P and temperature T. Since the end of the

18th century we knew these three macroscopic variables are connected by the phenomenological 'gas law':

$$PV = RT, \tag{2.1}$$

where R is the gas constant, a universal constant for all gases). The laws ruling gases were crucial for understanding how to convert heat to motive force during the industrial revolution. At a fundamental microscopic level a gas can be described by molecules of mass m moving with velocity v, obeying the known laws of particle mechanics (Newton's law, force = mass × acceleration). Given that there are $N_A = 6.02 \times 10^{23}$ (Avogadro's number) molecules in a mol of gas, it is clearly out of the question to solve the corresponding equations of motion explicitly. Therefore, physicists resort to a statistical physics formulation of particle mechanics, *statistical mechanics*, to derive these phenomenological results of thermodynamics from a probabilistic examination of the underlying microscopic systems. The approach of statistical mechanics, as its name implies, is to express the thermodynamic macroscopic variables as statistical averages of the microscopic variables of particle mechanics. James Clerk Maxwell (Maxwell 1867) assumed that: (1) the gas is composed of identical molecules moving in random directions, separated by distances that are large compared with their size and (2) the molecules undergo perfectly elastic collisions (no energy loss) with each other and with the walls of the container, but otherwise do not interact. With this hypothesis Maxwell was able to express both the pressure P and the temperature T of the gas in terms of the mean square velocity $\overline{v^2}$:

$$P = \frac{N_A}{3V} m\overline{v^2}, \tag{2.2}$$

$$T = \frac{N_A}{3R} m\overline{v^2}. \tag{2.3}$$

It is immediate to check that these two expressions lead to the gas law equation (2.1). Statistical mechanics thus explains the gas law by deriving this equation from the more fundamental microscopic laws of particle mechanics. Furthermore, statistical mechanics allows a microscopic interpretation for the temperature as a quantity proportional to the mean kinetic energy of molecule $\varepsilon_k = 1/2m\overline{v^2} = \overline{1/2mv^2}$.

Furthermore, Maxwell provided the probability velocity distribution:

$$f_{\text{Maxwell}}(v_x, v_y, v_z)\mathrm{d}v_x\, \mathrm{d}v_y\, \mathrm{d}v_z \propto \mathrm{e}^{-\frac{mv^2}{2k_B T}}\mathrm{d}v_x\, \mathrm{d}v_y\, \mathrm{d}v_z, \tag{2.4}$$

where $k_B \equiv R/N_A$.

The model used by Maxwell describes a *perfect* or *ideal* gas and is a reasonable approximation to a real gas, particularly in the limit of extreme dilution and high temperature. Such a simplified description, however, is not sufficiently precise to account for the behavior of gases at high densities. The Austrian physicist Ludwig Boltzmann later extended these ideas to real gases (with interacting molecules), liquids and solids providing a general statistical connection between the distribution of molecular energies and temperature.

2.1.2 Entropy and the Second Law of Thermodynamics: from the efficiency of heat engines and refrigerators to entropy as disorder

As we will see in the next subsection, a quantity which played an instrumental role in the development of statistical physics since Boltzmann is **entropy**. Let us briefly review in this subsection how this concept originated in thermodynamics in the 19th century connected to the problem of heat flow and became crucial to formulating the ***Second Law of Thermodynamics***. The ***First Law of Thermodynamics***, is basically a statement of conservation of energy. It distinguishes two kinds of transfer of energy, heat, Q, and thermodynamic work, W, and relates them to a function of a body's state, called internal energy U. So, in cases in which the system can exchange energy with its surroundings but not matter, the first law is written as $\Delta U = Q + W$, where ΔU denotes the change in the internal energy of such a system. The First Law of Thermodynamics makes no distinction between processes that occur spontaneously and those that do not. However, we find that only certain types of energy-conversion and energy-transfer processes actually take place. The Second Law of Thermodynamics establishes which processes do and which do not occur spontaneously in nature. This law has been expressed in many ways. For instance, heat always flows spontaneously from a hotter to a colder body. In fact, heat can never pass spontaneously from a colder to a warmer body without some other change. Its first formulation is Carnot's theorem which stated that:

> **The efficiency of conversion of heat to work in a heat engine has an upper limit.**
> Nicolas Léonard Sadi Carnot (1824)

The first rigorous definition of the second law based on the concept of entropy came from Rudolph Clausius, who defined entropy in 1865 (Clausius 1865). This definition requires that we introduce the concept of a ***reversible*** thermodynamic process. A reversible process is a process in which at all the intermediate states the system is always at (quasi-)equilibrium and such that the system and the environment can be restored to exactly the same initial states that they were in before the process occurred, if we go backward along the path of the process. The necessary condition for a reversible process is therefore a quasi-static variation. In fact it is quite easy to restore a system to its original state; the hard part is to have its environment restored to its original state at the same time. For example, if we have an ideal gas expanding into vacuum to twice its original volume, we can easily push it back with a piston and restore its temperature and pressure by removing some heat from the gas. The problem is that we cannot do it without changing something in its surroundings, such as dumping some heat there. Actually, a reversible process is an idealization that can never be precisely attained in the real world; real processes are irreversible because in practice energy dissipation is unavoidable (for example, due to friction between the piston and the container). But we can keep the system very close to equilibrium states and make the process nearly reversible. That's why we call a reversible process a quasi-equilibrium process.

Clausius considered an infinitesimal reversible process in which a system changes from one equilibrium state to another. If δQ_r is the amount of energy transferred by heat when the system follows this infinitesimal path, then there exists a function, called entropy S, such that the change in entropy dS is equal to this amount of energy for the reversible process divided by the absolute temperature of the system:

$$\mathrm{d}S = dQ_r/T. \tag{2.5}$$

Entropy, like internal energy, depends only on the *state* of the system, and its change between a final and an initial state $\Delta S \equiv S_f - S_i$, depends only on these two states. That is, entropy is a ***state function***. In contrast, heat, like work, is not a state function: the heat transferred during a thermodynamic process not only is a function of the end states of the process but also depends on the path followed in going from one state to another. For this reason, heat and work are called path functions or, in mathematical parlance, they are ***inexact differentials*** (and this is why we denote their infinitesimal variation by 'δ' instead of a 'd'). By equation (2.5), the change in entropy when a system moves between any two initial and final equilibrium states can be calculated as:

$$\Delta S = \int_i^f \mathrm{d}S = \int_i^f \frac{\delta Q_r}{T} \quad \text{(reversible path)}. \tag{2.6}$$

Therefore, although the change in entropy depends only on the initial and final thermodynamic states and is independent of the process connecting them (reversible or irreversible), its calculation requires a reversible path connecting the initial and final equilibrium states. This is because, according to equation (2.5), dS is only defined for reversible variations. Indeed, changes in entropy for real (irreversible) processes are calculated by devising a reversible process between the same two states and computing ΔS through equation (2.6) for the reversible process.

In a reversible thermodynamic process, by definition, the entropy of the system plus its surroundings, remains constant. For example, imagine an infinitesimal quasi-static compression of a gas in a cylinder which is in equilibrium with its surroundings at a temperature T. If there is no friction between the piston and the cylinder, by simply moving the piston infinitesimally in the opposite direction we can return the gas to its original conditions (volume and pressure) along the same path on a PV diagram without increasing the total entropy. Conversely, it can be shown that the change in entropy for a system and its surroundings is always positive for an irreversible process. In general, the total entropy always increases in an irreversible process. Since a system and its surroundings (in the limiting case, the rest of the entire Universe) form an isolated system, we can state the Second Law of Thermodynamics as follows:

The total entropy of an isolated system that undergoes a change can never decrease.
Rudolf Clausius (1865)

Therefore, either the total entropy either stays constant (reversible process), or it increases (irreversible process). Indeed, Clausius (1865) summarized the First and Second Laws of Thermodynamics in a sentence:

> ***'The energy of the universe is constant. The entropy of the universe tends to a maximum.'***

An equivalent way to formulate the Second Law is:

> **The values assumed by the extensive thermodynamic parameters in the absence of an internal constraint are those that maximize the entropy over the set of possible equilibrium states.**

In other words, in the absence of a constraint the system is free to select any one of a number of states, each of which might also be realized in the presence of a suitable constraint. The entropy of each of these constrained equilibrium states is definite, and the entropy is largest in some particular state of the set. In the absence of the constraint this state of maximum entropy is selected by the system (Callen 1985).

We have gained considerable insight about possible and impossible processes, but what is exactly the entropy really? In physics entropy is commonly associated with 'disorder'. What does this mean? We will adopt the Boltzmann's microscopically-based definition: disorder increases with the number of possible arrangements of molecules. In the words of Boltzmann (1896):

> 'From the standpoint of mechanics, any arrangement of molecules in the container is possible; in such an arrangement, the variables determining the motion of the molecules may have different average values in one part of the space filled by the gas than in another, where for example the density or mean velocity of a molecule may be larger in one half of the container than in the other, or more generally some finite part of the gas has different properties than another. Such a distribution will be called molar-ordered [*molar-geordnete*].'
>
> 'If the arrangement of the molecules also exhibits no regularities that vary from one finite region to another—if it is thus molar-disordered—...'

In the next subsections we will elaborate on the subject of this microscopic definition of entropy and how it led Boltzmann to combine the laws of mechanics and probability giving birth to statistical mechanics, but before that, let us illustrate this equivalence between entropy and disorder with a couple of examples. The first example we will consider is related with the three phases of water—ice, liquid and vapor—and transitions between them (figure 2.1). The second example is an adiabatic expansion of a gas.

Figure 2.1. A schematic plot showing the increasing disorder as we pass from solid to gas state. This Solid-liquid-gas imge has been obtained by the author from the Wikimedia website, where it is stated to have been released into the public domain. It is included within this article on that basis.

Example 1: Entropy of the different phases of water and change of entropy in melting

Ice, with atoms assembled in a regular crystal lattice displaying long-range order, is clearly the most ordered state. At the other extreme, gas, with its molecules moving randomly is the most disordered state. The molecules in the liquid phase are in an intermediate state of disorder; they are not arranged in a repeating three-dimensional array, like the ions in solids but, unlike the molecules in gases, the arrangement of the molecules in a liquid is not completely random. So the entropy is low for ice, medium for liquid and high for vapor. Hence, imagine we take an ice cube out of the refrigerator. As the ice cube absorbs heat from the hotter surroundings, which are at room temperature T_R, it melts. Assuming that the melting occurs so slowly that it can be considered a reversible process, the entropy gained by water by equation (2.6) is given by:

$$\Delta S = \frac{\int_i^f \delta Q_r}{T_R} = \frac{Q_r}{T_R}. \tag{2.7}$$

This increase in entropy is consistent with the fact that the ice transforms into the more disordered liquid state.

Example 2: Entropy change in an adiabatic expansion.

The second example is an adiabatic process, i.e., a process during which no energy enters or leaves the system by heat, that is $Q = 0$. Specifically, we will consider a gas in a thermally insulated container, as shown in figure 2.2. A wall separates the gas from an evacuated region. When the wall is removed, the gas expands freely into the vacuum. No heat is transferred to or from the gas because the container is insulated from its surroundings. (In addition, because the gas does not exert a force through a distance on the surroundings, it does no work on the surroundings as it expands.)

Equation (2.5) might give you the wrong idea that entropy change is exclusively related with heat transfer, and, since $Q = 0$ there is no change in entropy. But this is wrong. Clearly this adiabatic expansion is an irreversible process; the gas will not spontaneously come back to occupy just its original smaller volume. Even though there has been no heat

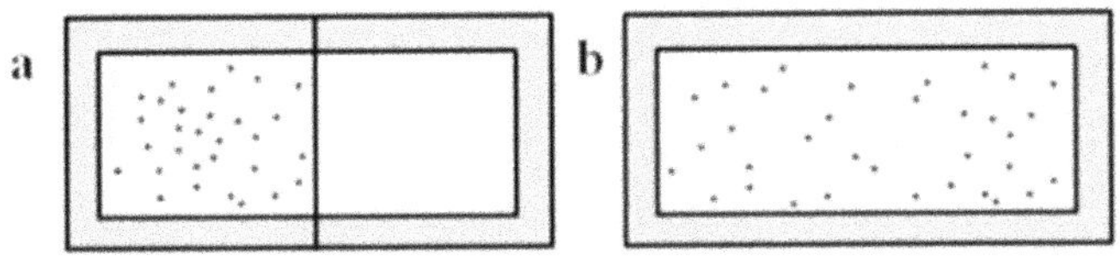

Figure 2.2. Adiabatic expansion. A thermally isolated box. In (a) the gas is confined in the left compartment. Then the partition is removed and the gas expands freely until it occupies all the volume in (b).

transfer in this process, it is straightforward to show that the entropy of the gas has increased in this process without heat flow. To do this by applying equation (2.6), we cannot use $Q = 0$, the value for the irreversible process. Rather, as mentioned, we must find an equivalent reversible path that shares the same initial and final states and find the associated Q_r. A simple choice is an isothermal, reversible expansion in which the gas pushes slowly against a piston while energy enters the gas in the form of heat from a heat reservoir to hold the temperature constant. For such isothermal process, the First Law of Thermodynamics (i.e., the conservation of energy) implies that the heat absorbed by the gas, Q_r, must be equal to the work W it has done during the expansion from the initial volume, V_i, to the final volume, V_f. Thus we have:

$$Q_r = W = \int_{V_i}^{V_f} P\, \mathrm{d}V. \tag{2.8}$$

Then, by equation (2.1) we have:

$$Q_r = \int_{V_i}^{V_f} \frac{RT}{V} \mathrm{d}V = RT \ln \frac{V_f}{V_i}. \tag{2.9}$$

Because T is constant in this process, equations (2.6) and (2.9) give:

$$\Delta S = \frac{Q_r}{T} = R \ln \frac{V_f}{V_i}. \tag{2.10}$$

As $V_f > V_i$ we conclude that ΔS is positive. By the same token the disorder also has grown since the molecules end with many more possible arrangements when occupying a final larger volume. This indicates that both the entropy and the disorder of the gas increase as a result of the irreversible, adiabatic expansion.

2.1.3 Boltzmann and the statistical character of the Second Law of Thermodynamics

A general procedure to obtain the probability distribution for the particle energies, devised by Boltzmann in 1877 (Boltzmann 1877), is to take the distribution that maximizes the entropy of the gas. Boltzmann imagined gas molecules occupying s different energetic states $i = 1,2,3,\ldots,s$ of definite discrete energy[1] ε_i. Thus the total energy of the gas E would be equal to

[1] Boltzmann found it easier to think about discrete energies than continuous ones. After the advent of quantum mechanics, at the beginning of the 20th century, this seems a well justified choice. In fact, it seems that Boltzmann was a pioneer of quantum mechanical ideas; he is credited to have said at the Halle conference in 1891: 'I see no reason why energy shouldn't also be regarded as divided atomically.'

$$E = \sum_{i=1}^{s} n_i \varepsilon_i, \tag{2.11}$$

where n_i denotes the number of molecules with energy ε_i. Actually equation (2.11) is an approximate formula that is valid only when the interaction energy between molecules is negligible (otherwise we would have to include in the sum these interaction energies). This is the case for a perfect or ideal gas defined as a gas for which the intermolecular potential energy of interaction is negligible. Cleary, if N is the number of gas molecules, we would also have:

$$N = \sum_{i=1}^{s} n_i. \tag{2.12}$$

Now any set $\{n_i\}$ of occupation numbers for which E, N agree with the given information, represents a possible distribution, compatible with all that is specified. At this point it is convenient to distinguish between a **microscopic state** or ***microstate*** and **macroscopic state** or ***macrostate***. The microstate i is the actual set of velocities, and positions, of all the molecules. Thus the definition of a microstate requires the specification of $3N$ position coordinates (three per molecule) $q_1,q_2,\dots,q_{3N}$ and $3N$ momentum components $p_1,p_2,\dots,p_{3N}$ (also three per molecule). Geometrically, this set of $6N$ quantities may be regarded as a point (q_i,p_i) in a space of $6N$ dimensions, called the ***phase space*** and the phase point (q_i,p_i) as a *representative point*, of the given system.

Thus Boltzmann imagined each gas molecule occupying a small six-dimensional volumetric cell i ($i = 1, 2, 3, \dots, s$) of phase space, with a well-defined energy ε_i. On the other hand, a **macroscopic state** or ***macrostate*** is characterized by a vector of occupation numbers $\vec{n} = [n_1,n_2,\dots]$, i.e., n_1 molecules in the molecular energy level ε_1, n_2 molecules in the ε_2, and so on so forth. Since we are assuming that molecules are indistinguishable, there are many microstates corresponding to a given macrostate of the gas: permutations of indistinguishable molecules between different single molecular energetic levels i of a microstate correspond to the same macrostate with total energy E, number of particles N total volume V. As is well known, the number of microstates corresponding to a given macrostate, or *multiplicity* $\mathcal{W}$, is given by the multinomial formula (Pathria and Beale 2011):

$$\mathcal{W}(E, V, N) = \frac{N!}{n_1! n_2! \cdots n_s!}. \tag{2.13}$$

For a macroscopic system (i.e., N of the order of N_A) $\mathcal{W}$ is in general huge. The number of macrostates, which is smaller than the number of microstates, is still very large (box 2.1 illustrates the growing of both microstates and macrostates with the number N of particles).

It turns out that the multiplicity $W(E,V,N)$ is a central ingredient in Boltzmann's derivation of statistical mechanics. In fact, Boltzmann proposed the Fundamental Postulate of Statistical Mechanics which states that:

Box 2.1. The number of microstates and macrostates for two coins

To illustrate the counting of states imagine a very simplified system in which each molecule or particle has only two possible accessible states, and thus can be represented as a coin with states either head (H) or tail (T).

Let's start with just two coins. If we toss up two coins, the possible outcomes are:

coin 1	coin 2
H	H
H	T
T	H
T	T

The four possible outcomes of this set of two coins with two states H and T are the ***microstates***: HH, HT, TH, TT.

If the two coins are exactly the same; i.e., ***indistinguishable***, then the two microstates HT and TH constitute one ***macrostate***, in such a way we have:

1 Head macrostate: $\mathcal{W}_1 = 2$ microstates (in agreement with equation (2.3): $2!/1!\,1! = 2/1 = 2$)

2 Heads macrostate: $\mathcal{W}_2 = 1$ microstate (in agreement with equation (2.3): $2!/2!\,0! = 2/2 = 1$)

0 Heads macrostate: $\mathcal{W}_0 = 1$ microstate (in agreement with equation (2.3): $2!/0!\,2! = 2/2 = 1$)

If we denote the number of distinguishable microstates by m and the number of indistinguishable macrostates by M, we have:

$m = \mathcal{W}_0 + \mathcal{W}_1 + \mathcal{W}_2 = 4$ (alternatively, the number of microstates can be obtained as $m = 2^2 = 4$), $M = 3$.

If we now consider $N = 100$ coins, the total number of microstates is $m = 2^{100} \approx 1.2677 \times 10^{30}$ and the number of macrostates (0 H, 1 H, 2 H, ..., 100 H) is $M = 101$.

Therefore we observe an important result: for binary particles, $M \sim N$ and $m \sim 10^{N/3.3}$.

All microstates consistent with a given macrostate of a system are equally likely[2]. This means that when we observe a particular macrostate in an experiment, we are averaging over all the microstates that it comprises, giving equal weight (often called the *a priori* probability) to each of them. In classical statistical mechanics, this assumption follows from the ***ergodic hypothesis*** or ***ergodic theorem***. The ergodic

[2] Assuming there are no other constraints than E, V, N. Adding other constraints would restrict the system accessible states. For example, imagine we divide the volume with a wall into a left and a right compartment with volumes V_L and V_R. And, we prepare the system in a state in which all the gas molecules are in the left compartment. Then, if the wall is removed the new equilibrium corresponds to a situation in which the gas occupies the whole volume V. This is because the number of accessible states for each molecule suddenly doubled. In other words, when we remove constraints the number of accessible states grows and thus the equilibrium state changes to one with a larger multiplicity.

hypothesis basically states that a macroscopic system passes through all the W possible microstates, spending on average an equal amount of time in each. Thus, a time average is equivalent to an average over the microstates. This equivalence has been a controversial issue regarding the foundations of statistical mechanics. Most proofs are highly abstract and far removed from physics. Indeed, there are examples of macroscopic systems that violate the literal form of the ergodicity and a proof of the ergodic theorem under fairly general conditions is lacking. For a review about the status of the ergodic theorem we refer the reader to the review by C C Moore (2015).

Hence, out of the millions of such possible macrostates, which is the one most likely to be realized for a system at equilibrium? Boltzmann's answer was that the 'most probable' equilibrium macrostate is the one that can be realized in the greatest number of ways, i.e., the one with greatest multiplicity $\mathcal{W}$. Thus if we have a macroscopic system in a macrostate with energy E, volume V and number of particles N the probability of this macrostate is proportional to the number of microstates corresponding to it $\mathcal{W}(E, V, N)$:

$$P(E,\ V,\ N) \propto W(E,\ V,\ N). \tag{2.14}$$

And, we are saying that $\mathcal{W}(E, V, N)$ is maximum for the equilibrium state.

Now, according to the Second Law of Thermodynamics such equilibrium state also maximizes the thermodynamic entropy S. Boltzmann thus reasoned that there must be a simple relationship between $\mathcal{W}(E, V, N)$ and S. Since $\mathcal{W}(E, V, N)$, for two isolated systems is multiplicative, it follows that the logarithm of $\mathcal{W}(E, V, N)$ is additive. In other words $\ln \mathcal{W}(E, V, N)$, like the entropy, is an extensive quantity. Therefore, Boltzmann proposed the simplest possible relationship, i.e., $S = k_B \ln W$, where k_B is the so-called Boltzmann's constant (connected with the gas constant R by $k_B = R/N_a$). Notice that this formalizes his definition of **disorder as an increase of permissible or accessible microstates**.

In this way, the Second Law of Thermodynamics arises from the fully reversible microscopic laws of mechanics by the laws of probability. That is, the entropy of a state measures its probability, so **the entropy increases because systems evolve from less probable states to most probable states.**

2.1.4 Boltzmann–Gibbs maximum entropy approach to statistical mechanics

Taking the logarithm of equation (2.14) we get:

$$S = k_B \ln \mathcal{W} = k_B(\ln N! - \ln(n_1!n_2!\cdots n_s!)) = k_B\left(\ln N! - \sum_{i=1}^{s} \ln n_i!\right). \tag{2.15}$$

Since the n_i are large we can approximate the factorial by Stirling's formula ($\ln n! \approx n \ln n - n$), obtaining the ***Gibbs entropy formula***:

$$S \approx -k_B N \sum_{i=1}^{s} (n_i/N)\ln(n_i/N) = -k_B N \sum_{i=1}^{s} p_i \ln p_i, \tag{2.16}$$

where $p_i = n_i/N$ is the probability that a molecule has energy ε_i (figure 2.3).

Figure 2.3. The fathers of statistical mechanics. (Left) this James Clerk Maxwell image has been obtained by the author from the Wikimedia website, where it is stated to have been released into the public domain. It is included within this article on that basis. (Middle) this Ludwig Boltzmann image has been obtained by the author from the Wikimedia website, where it is stated to have been released into the public domain. It is included within this article on that basis. (Right) Gibbs (1907), copyright 1907, Nature Publishing Group, with permission of Springer.

Boltzmann then argued that the occupation probabilities p_i of the most probable state at equilibrium are those that maximize the entropy, $S = -k_B N \sum_{i=1}^{s} p_i \ln p_i$, and which also satisfy the constraint on the total particle number N (2.12) and the constraint on the total energy E (2.11). These two constraints can be written in terms of occupation probabilities p_i as:

$$\begin{cases} \sum_{i=1}^{s} p_i = 1, & \text{(2.17a)} \\ \sum_{i=1}^{s} p_i \varepsilon_i = \bar{\varepsilon}, & \text{(2.17b)} \end{cases}$$

where $\bar{\varepsilon} = E/N$ is the average energy (per particle). The state of equilibrium is computed by maximizing the entropy subject to the constraints. The mathematical procedure to do this is the ***method of Lagrange multipliers***, which is equivalent to maximizing the variational function $v(\{p_j\})$ defined as:

$$\mathbf{v}\left(\{p_j\}\right) \equiv \frac{S\left(\{p_j\}\right)}{N k_B} - \lambda_0 \left(\sum_{j=1}^{s} p_j - 1 \right) - \lambda_1 \left(\sum_{j=1}^{s} p_j \varepsilon_j - \bar{\varepsilon} \right), \tag{2.18}$$

where λ_0 and λ_1 are new variables called *Lagrange multipliers*. This variational function has $s + 2$ unknowns: s different p_j's and the two Lagrange multipliers. The maximum condition implies varying equation (2.18) with respect to each p_i and equating these variations to zero. This produces a set of equations that uniquely determines the $s + 2$ unknowns. Let's do this; the variation of equation (2.18) with respect to each p_i yields:

$$\frac{\mathrm{dv}}{\mathrm{d}p_i} = -\ln p_i - 1 - \lambda_0 - \lambda_1 \varepsilon_i, \tag{2.19}$$

and setting this to zero gives that the p_i values that maximize the entropy subject to the constraints equation (2.17) are given by:

$$p_i^* = \mathrm{e}^{-1-\lambda_0-\lambda_1\varepsilon_i}. \tag{2.20}$$

To obtain λ_0 we use the normalization constraint (2.17a), and summing over i on both sides of equation (2.20) we get: $\mathrm{e}^{1+\lambda_0} = Z_1$, where $Z_1 \equiv \sum_{i=1}^{S} \mathrm{e}^{-\lambda_1\varepsilon_i}$ is called the molecular or single-particle *partition function* (Glazer and Wark 2006); the Z stands for the German *Zustandssumme* ('sum of state'). Thus (2.20) can be re-written as:

$$p_i^* = \frac{\mathrm{e}^{-\lambda_1\varepsilon_i}}{Z_1}. \tag{2.20'}$$

In turn, λ_1 can be identified by making contact with thermodynamics. Indeed, the First Law of Thermodynamics (principle of energy conservation) applied for a reversible process over a gas that keeps the volume constant allows connecting variations of entropy and energy at a given temperature T as $\mathrm{d}E = T\,\mathrm{d}S$. Using equations (2.11) and (2.15), this relation can be written as:

$$N\,\mathrm{d}\left(\sum_{i=1}^{S} p_i\varepsilon_i\right) = -Nk_{\mathrm{B}}T\,\mathrm{d}\left(\sum_{i=1}^{S} p_i \ln p_i\right). \tag{2.21}$$

The differential variation of $\sum_{i=1}^{S} p_i\varepsilon_i$ is given by:

$$\mathrm{d}\left(\sum_{i=1}^{S} p_i\varepsilon_i\right) = \sum_{i=1}^{S}\left(p_i\,\mathrm{d}\varepsilon_i + \varepsilon_i\,\mathrm{d}p_i\right).$$

Nevertheless, since the volume is fixed there is no work over the gas and therefore the energy levels ε_i are kept constant, i.e., $\mathrm{d}\varepsilon_i = 0$ and so the above equation simplifies to:

$$\mathrm{d}\left(\sum_{i=1}^{S} p_i\varepsilon_i\right) = \sum_{i=1}^{S} \varepsilon_i\,\mathrm{d}p_i.$$

The differential variation of $\sum_{i=1}^{S} p_i \ln p_i$ in turn produces:

$$\mathrm{d}\left(\sum_{i=1}^{S} p_i \ln p_i\right) = \left(\sum_{i=1}^{S} \mathrm{d}p_i \ln p_i + p_i 1/p_i\,\mathrm{d}p_i\right) = \sum_{i=1}^{S}(\ln p_i + 1)\mathrm{d}p_i. \tag{2.22}$$

Substituting equation (2.20′) into equation (2.22) we get:

$$\mathrm{d}\left(\sum_{i=1}^{S} p_i \ln p_i\right) = -\lambda_1\sum_{i=1}^{S}\varepsilon_i\,\mathrm{d}p_i + (1 - \ln z)\sum_{i=1}^{S}\mathrm{d}p_i = -\lambda_1\sum_{i=1}^{S}\varepsilon_i\,\mathrm{d}p_i, \tag{2.23}$$

where the last equality in equation (2.23) holds because $\sum_{i=1}^{S}\mathrm{d}p_i = \mathrm{d}\sum_{i=1}^{S}p_i$ and, by equation (2.17a) $\mathrm{d}\sum_{i=1}^{S}p_i = 0$. Then, substituting equations (2.22) and (2.23) into equation (2.21) we get

$$\lambda_1 = \frac{1}{k_B T}, \tag{2.24}$$

and thus equation (2.20′) becomes the Boltzmann probability distribution for a single molecule:

$$p_i^* = \frac{e^{-\frac{\varepsilon_i}{k_B T}}}{Z_1}. \tag{2.25}$$

Notice that equation (2.25) tells us that:

(i) The relevant scale for molecular energies is $k_B T = RT/N_a$, which, as we have seen, is (aside from a multiplicative factor) the mean molecular kinetic energy.

(ii) The probability of occupation of a state of energy ε_i increases when the *mean* molecular kinetic energy grows (i.e., when the temperature grows).

Notice that Boltzmann's distribution equation (2.25), in the case of an ideal gas of free (non-interacting) molecules with $\varepsilon_i = 1/2mv_i^2$, reduces to Maxwell's velocity distribution law, equation (2.4). Alternatively, for a gas in a gravitational field, $\varepsilon_i = 1/2mv_i^2 + mgh$, where h stands for the height above sea level, equation (2.25) gives the usual *barometric formula* for decrease of the atmospheric density with height: $\rho(h) = \rho(0)\exp(-mgh)$.

Although in the derivation of the Boltzmann probability distribution equation (2.25) we have assumed that the interaction energy between molecules is negligible (equation (2.11)), Josiah Willard Gibbs (1902) later showed that it holds in general when interactions between molecules are non-negligible. Gibbs resorted in his derivation to the concept of ***statistical ensemble***, an idealization consisting of a large number of virtual copies (sometimes infinitely many) of a system, considered all at once, each of which represents a possible state that the real system might be in. In other words, a statistical ensemble is equivalent to a probability distribution for the state of the system. Gibbs argued that, at equilibrium, the probability distribution of states in the phase space of all coordinates and momenta, $P^*(q_i, p_i)$, must depend on these only through conserved quantities such as energy $E(q_i, p_i)$ (Landau and Lifshitz 1980). Note that E denotes the energy of a system of the N particles as a whole in contrast to ε_i, used earlier, that denote single-particle energy levels. The energy levels for the whole system of N particles and the molecular energy levels are connected by equation (2.11). The equilibrium probability distribution, for a gas in thermodynamic equilibrium at a temperature T and occupying a volume V, can be written as:

$$P^*(q_i, p_i) = \frac{e^{-\frac{E(q_i, p_i)}{k_B T}}}{Z_N[T, V]}, \tag{2.26}$$

with $Z_N[T, V]$, known as the **canonical partition function** given by

$$Z_N[T, V] \equiv \int \cdots \int e^{-\frac{E(q_i, p_i)}{k_B T}} \frac{dq_1 \cdots dp_{3N}}{h^{3N}}, \tag{2.27}$$

where h in the normalization factor for the classical state element of phase space, with dimension of $[qp]$ making thus the partition function non-dimensional. For the case of a macroscopic system with discrete energy levels, E_s ($s = 1,2,\ldots$), equations (2.26) and (2.27) become:

$$P_s^* = \frac{e^{-\frac{E_s}{k_B T}}}{Z_N[T, V]}, \tag{2.26$'$}$$

and

$$Z_N[T, V] \equiv \sum_{s=1} e^{-\frac{E_s}{k_B T}}. \tag{2.27$'$}$$

The probability distribution (2.26) (or (2.26′)) defines the so-called ***canonical ensemble***, which corresponds to physical systems in contact with a heat reservoir or heat bath that keeps the temperature of the system fixed to T.

$Z_N[T, V]$, by virtue of equation (2.11) can be written in terms of the single-particle partition function Z_1 as:

$$Z_N[T, V] = Z_1[T, V]^N. \tag{2.28}$$

Suppose that we measure the value of a dynamical variable $A(q_i, p_i)$ of some macroscopic body at time t, and call the corresponding outcome $A_{\text{meas}}(t)$. This measurement is not an instantaneous process. Suppose we want to measure the pressure P of a tire with a manometer. If we observe the time evolution of P, we will find that P fluctuates as a result of particles bouncing around in the surface area of the column of liquid until it reaches a stationary value. The typical time τ we have to wait is of the order of seconds. Thus, what we obtain as a result of this measurement is not the instantaneous value $A(t)$ assumed by the dynamical variable, rather it is a **time average** of $A(t)$. So, a proper expression for $A_{\text{meas}}(t)$ is

$$A_{\text{meas}}(t) = \frac{1}{\tau} \int_t^{t+\tau} A(q_i(t'), p_i(t'))\, dt'. \tag{2.29}$$

On the other hand, the characteristic time scale of air molecular motion τ' can be estimated using a back-of-the-envelope calculation: the velocity of molecules in air is of the order of the velocity of sound (330 m s^{-1}) and they are bouncing around between scales of few (say 10) angstroms (10^{-10} m). So $\tau' \sim 10^{-9}$ m/3.3×10^3 m s^{-1} ~ 10^{-12} s. Since $\tau \gg \tau'$, for all practical purposes, we can replace equation (2.29) by:

$$A_{\text{meas}}(t) = \lim_{\tau \to \infty} \frac{1}{\tau} \int_t^{t+\tau} A(q_i(t'), p_i(t'))\, dt'. \tag{2.30}$$

The probability distribution equation (2.26) can be used to compute the expected values of dynamical variables $A(q_i, p_i)$ as:

$$\langle A \rangle \equiv \int \cdots \int A(q_i, p_i) \mathrm{e}^{-\frac{E(q_i, p_i)}{k_B T}} \frac{\mathrm{d}q_1 \cdots \mathrm{d}p_{3N}}{h^{3N}}, \tag{2.31}$$

or, in the case of discrete energies, as:

$$\langle A \rangle \equiv \sum_{s=1} A_s \mathrm{e}^{-\frac{E_s}{k_B T}}. \tag{2.31}$$

We have seen in section 2.1.3 that the ergodic hypothesis guarantees that a macroscopic system passes through all its possible microstates, spending on average an equal amount of time in each. Thus, this hypothesis also implies the equivalence between time and ensemble averages, i.e.,

$$\langle A \rangle = A_{\text{meas}}(t) \tag{2.32}$$

In other words:

Ensemble average of a quantity A = *Time average of a* = Measured value for A

2.1.5 All is in the partition function

The partition function is the central object of statistical mechanics: if we are able to compute the partition function all the thermodynamic quantities follow straightforwardly. Indeed the connection with thermodynamics is provided by the entropy function, in the so-called ***thermodynamic limit***, where both E and $N \to \infty$ for a fixed value of the ratio E/N. Let us thus start by substituting the Boltzmann probability distribution for a single molecule, equation (2.25), into the Gibbs formula (2.16), so that the entropy S can be written as:

$$S = Nk_B \left[\ln Z_1[T, V] + T \frac{\partial \ln Z_1[T, V]}{\partial T} \right]. \tag{2.33}$$

And, by (2.28) we get:

$$S = k_B \ln Z_N[T, V] + k_B T \frac{\partial \ln Z_N[T, V]}{\partial T}. \tag{2.34}$$

Next, it is immediate to realize that equating the internal energy U of the system with the ensemble mean energy $\langle E \rangle \equiv \sum_{s=1} P_s E_s$ (equation (2.31′) U is obtained from the partition function equation (2.27) as

$$U = \langle E \rangle \equiv k_B T^2 \frac{\partial \ln Z_N[T, V]}{\partial T}. \tag{2.35}$$

This internal energy is of a system of particular thermodynamic interest; it can be expressed in terms of pairs of conjugate variables such as temperature T and entropy

S or volume V and pressure P. The product of two quantities that are conjugate has units of energy. So, once we have U, we can easily compute all the other thermodynamic variables. In particular, the thermodynamic conjugate variable to V, P, is obtained as derivatives of U with respect to these variables. Thus,

$$P = -\frac{\partial U}{\partial V} = -k_B T^2 \frac{\partial^2 \ln Z_N[T,\ V]}{\partial T \partial V}. \tag{2.36}$$

With illustrative purposes we will present the calculation of the canonical partition function for a very popular model in statistical mechanics textbooks, the Ising model, invented by the physicist Wilhelm Lenz (1920) as a mathematical model of ferromagnetism. It is named after Ernst Ising, a student of Lenz, who solved this model in one spatial dimension in his 1924 thesis (Ising 1925). The original Ising model involves N magnetic dipoles fixed on the nodes of a regular spatial lattice. These dipoles with only two possible orientations, up and down, are represented by spin variables or 'spins' $\sigma_i = -1$ or $+1$; each spin interacts only with its lattice nearest neighbors. The total energy of the system of spins is the sum of two terms. First, we have the interaction energy ε_{ij} between two spins or particles i and j, which depends on their states, and can be written as:

$$\varepsilon_{ij} = -J_{ij}\sigma_i\sigma_j = \begin{cases} -J_{ij} & \text{for } \sigma_i = \sigma_j, \\ +J_{ij} & \text{for } \sigma_i = -\sigma_j, \end{cases} \tag{2.37}$$

with the matrix J_{ij} defined as $J_{ij} = \begin{cases} J & \text{if nodes } i \text{ and } j \text{ are nearest neighbors,} \\ 0 & \text{otherwise.} \end{cases}$

Second, there is a term for the energy corresponding to the interactions of the magnetic dipole moments with an external magnetic field h, which is given by:

$$\varepsilon_i = -h\sigma_i = \begin{cases} -h & \text{for } \sigma_i = 1, \\ +h & \text{for } \sigma_i = -1. \end{cases} \tag{2.38}$$

Therefore, the total energy of the Ising model can be written as:

$$E = -J\sum_{<ij>} \sigma_i\sigma_j - h\sum_{i=1}^{N} \sigma_i, \tag{2.39}$$

where the brackets < > for the sum indices in the first sum denote that sites i and j are nearest neighbors. The canonical partition function $Z_N[T,h]$ (notice that the volume V, a relevant variable for a gas, has been replaced by a variable the external magnetic field h of importance for a magnetic material), is then given as a sum over all the spin configurations $\{\sigma_1, \sigma_2, \ldots, \sigma_N\}$:

$$Z_N^{\text{Ising}}[T,\ h] \equiv \sum_{\{\sigma_1,\sigma_2,\ldots,\sigma_N\}} \exp\left[\frac{J\sum_{<ij>} \sigma_i\sigma_j + h\sum_{i=1}^{N} \sigma_i}{k_B T}\right]. \tag{2.40}$$

Similarly to the internal energy, the magnetization M of the system coincides with the expected value of the sum of all magnetic dipole moments $M = \sum_{s=1} P_s M_s = \sum_{s=1} P_s \left[\sum_{i=1}^{N} \sigma_i \right]_s$. Therefore, in a completely analogous way to that with which we obtained the pressure for a gas, equation (2.36), we can obtain M as:

$$M = k_{\mathrm{B}} T \frac{\partial \ln Z_N^{\mathrm{Ising}}[T, h]}{\partial h}. \tag{2.41}$$

Indeed, a wide variety of physico-chemical systems that undergo phase transitions can be represented, to varying degrees of accuracy, by an array of lattice sites, with only nearest-neighbor interaction. This simple setting turns out to be good enough to provide a unified, theoretical basis for understanding a variety of phenomena such as ferromagnetism and antiferromagnetism, gas–liquid and liquid–solid transitions, order–disorder transitions in alloys, phase separation in binary solutions, and as a lattice gas model—i.e., a statistical model for the motion of atoms. Furthermore, the Ising model has been used in a variety of fields other than physics.

Since one of the main goals in this book is to provide conceptual and methodological bridges between physics, biology and economics we briefly review some applications of this model in the latter two scientific areas. For example, in biological systems, Ising-inspired models have been used to understand a range of binding behaviors, from DNA compaction (Vtyurina *et al* 2016) or the binding of ligands to receptors in the cell surface (Shi and Duke 1998) to the flagellar motor of bacteria (Bai *et al* 2010). A well-known application to neuroscience is as a minimal model of a neural network which is capable of learning (Hopfield 1982). That is, the activity of neurons in the brain can be modeled statistically, with two possible alternative binary states for each neuron: either active or inactive. The Ising model also has had several applications in social and financial systems (Sornette 2014). For example, each firm is connected to some other firms; companies can buy each other's products as intermediate goods or services. The Ising model allows us to describe the dynamics of such a network of firms and corporations with a focus on the role of social interactions in economic behavior (Brock and Durlauf 2001). The Ising framework has also been used to model changes in economic opinions (Hohnisch *et al* 2005) and the connected emergence of herding behavior, leading to investment bubbles and market crashes. The Ising model has been demonstrated to be useful to approach problems like urban segregation, where agents of two different 'colors' self-organize in spatial domains (Stauffer 2008).

Box 2.2 presents the calculation of the partition function for the simplest one-dimensional Ising model, which is quite straightforward. A one-dimensional schematic representation is shown in figure 2.4 Each spin interacts only with its two nearest neighbors; the ones to its left and to its right. For example, the red spin interacts with the two blue spins.

Unfortunately, the one-dimensional Ising model is too simple and exhibits no really interesting phenomena like a phase transition from the ferromagnetic ordered state to the paramagnetic disordered. The two-dimensional Ising model, in which the

Box 2.2. Exact solution of the Ising model in one dimension

Here we compute the canonical partition function for the Ising model in one dimension as a function of the thermodynamic parameters T and h. Next, we obtain the internal energy.

A simplifying condition for computing the partition function is to assume ***periodic boundary conditions*** in such a way that the Nth spin becomes a neighbor of the first. This avoids finite size effects at the boundaries, and it is equivalent to replace the straight open chain by a circle as shown in figure 2.4(b). Therefore, equation (2.39) for the energy can be written as:

$$E = -J\sum_{i=1}^{N}\sigma_i\sigma_{i+1} - h\sum_{i=1}^{N}\sigma_i \tag{i}$$

The canonical partition function $Z_N^{\text{Ising}}[T, h]$ then becomes:

$$Z_N^{\text{Ising}}[T, h] \equiv \sum_{\{\sigma_1,\sigma_2,\ldots,\sigma_N\}} \exp\left[\frac{J}{k_{\text{B}}T}\sum_{i=1}^{N}\sigma_i\sigma_{i+1} + \frac{h}{2k_{\text{B}}T}\sum_{i=1}^{N}(\sigma_i + \sigma_{i+1})\right], \tag{ii}$$

where the second term of the energy has been rearranged to take advantage of a more symmetric form, which allows us to rewrite:

$$Z_N^{\text{Ising}}[T, h] \equiv \sum_{\{\sigma_1,\sigma_2,\ldots,\sigma_N\}} \prod_{i=1}^{N} T(i, i+1), \tag{iii}$$

with

$$T(i, i+1) = \exp\left[\frac{J}{k_{\text{B}}T}\sum_{i=1}^{N}\sigma_i\sigma_{i+1} + \frac{h}{2k_{\text{B}}T}\sum_{i=1}^{N}(\sigma_i + \sigma_{i+1})\right]. \tag{iv}$$

$T(i,i+1)$ can be written as a 2×2 **transfer matrix**, whose indices are the spin variables, $\sigma_i = \pm 1$ and $\sigma_{i+1} = \pm 1$, defined by

$$\mathbf{T} = \begin{bmatrix} \exp(K + L) & \exp(-K) \\ \exp(-K) & \exp(K - L) \end{bmatrix}, \tag{v}$$

with $K = J/k_{\text{B}}T$, $L = h/k_{\text{B}}T$. Using matrix algebra we see that $Z_N^{\text{Ising}}[T, h]$ is a trace of a product of N identical transfer matrices, i.e.,

$$Z_N^{\text{Ising}}[T, h] = \text{tr}(\mathbf{T}^N). \tag{vi}$$

The transfer matrix is symmetric and thus it can be diagonalized by a unitary transformation $\mathbf{U}^{-1} = \mathbf{U}^{\dagger}$ (where $\dagger$ means transpose and conjugate):

$$\mathbf{UTU}^{-1} = \mathbf{D}, \tag{vii}$$

where $\mathbf{D}$ is a diagonal matrix. Therefore, the canonical partition function can be written in terms of the eigenvalues of the transfer matrix,

$$Z_N^{\text{Ising}}[T, h] = \text{tr}[(\mathbf{UTU}^{-1})^N] = \text{tr}[\mathbf{D}^N] = \lambda_1^N + \lambda_2^N = \lambda_1^N\left[1 + \left(\frac{\lambda_2}{\lambda_1}\right)^N\right], \tag{viii}$$

where the eigenvalues λ_1 and λ_2 are the roots of the secular determinant $\det(\mathbf{T} - \lambda\mathbf{I})$, i.e.,

$$\lambda_{1,2} = e^K \cosh L \pm \sqrt{e^{2K}\cosh^2 L - 2\sinh(2K)}\,. \tag{ix}$$

Notice that these eigenvalues are always positive, and that $\lambda_1 > \lambda_2$ (except at the trivial point $T = h = 0$). Thus, we arrive in the thermodynamic limit, $N \to \infty$, to the desired expression of the canonical partition function:

$$Z_N^{\text{Ising}}[T, h] = \left(e^K \cosh L + \sqrt{e^{2K}\cosh^2 L - 2\sinh(2K)}\,\right)^N \qquad (\text{for } N \to \infty). \tag{x}$$

Just for simplifying the algebra let us consider the case with no external magnetic field, h = 0, so that (x) reduces to

$$Z_N^{\text{Ising}}[T, 0] \to (2\cosh K)^N \qquad (\text{for } N \to \infty). \tag{xi}$$

Therefore, using equation (2.35) we obtain the internal energy as:

$$\mathrm{U} = -Nk_BT\tanh(J/k_BT) \tag{xii}$$

And, similarly we can obtain all the thermodynamics quantities of interest.

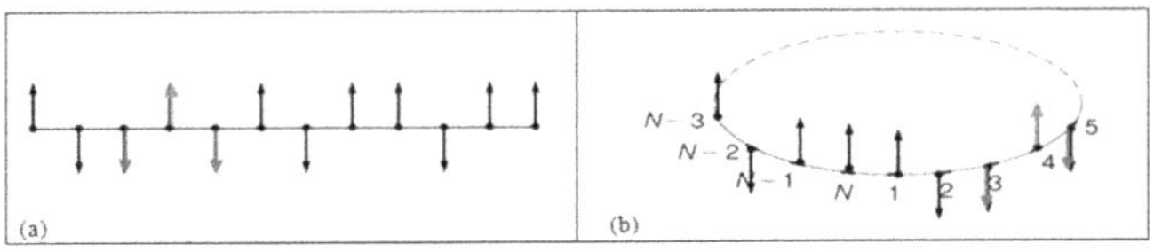

Figure 2.4. A one-dimensional lattice Ising model. Arrows represent microscopic magnetic dipoles that can point either upward or downward. (a) Open chain, (b) circular chain, corresponding to periodic boundary conditions (see box 2.2).

spins are arranged on a grid in a plane, exhibits such critical phenomena. However, the calculation of the partition function in two dimensions (Onsager 1944) is much more cumbersome than in one dimension. The model still defies exact solution if the spins form a three-dimensional lattice.

2.2 MaxEnt derivation of statistical mechanics

Remarkably, Boltzmann's statistical formulation didn't take into account the millions of intricate dynamical details of the motion of molecules, the only dynamic information that was included is that the total energy is conserved. In addition Boltzmann assumed ***ergodicity***, i.e., that all accessible microstates are equiprobable over a long period of time. However the fact that it is enough to predict the correct spatial and velocity distribution of the molecules shows that those microscopic details are actually irrelevant to the predictions (they cancel out when performing averages).

In the this section we will discuss the generalization of the Boltzmann's method to a more abstract and general principle of MaxEnt, for inferring probability distributions from limited data. This alternative derivation of the statistical mechanics is based on the identification of the Shannon and Boltzmann entropies.

2.2.1 The equivalence between Boltzmann's and Shannon's entropy

One of the main lessons of chapter 1 was that the information entropy coincides with the missing information (MI). This MI is the minimum number of binary questions we need to ask, knowing the macroscopic configuration of a system, to know its particular microscopic configuration. Specifically, we have seen that MI = $\log(W)$. Therefore, by virtue of equation (2.15), we have that

$$S \propto \mathrm{MI}. \tag{2.42}$$

That is, **thermodynamic entropy, from a statistical mechanics point of view, is basically equivalent to Shannon's information entropy**:

$$S = k_B H. \tag{2.43}$$

Entropy can thus be regarded as a measure of ignorance. When it is known only that a system is in a given macrostate, the entropy of the macrostate measures the degree of ignorance the microstate is in by counting the number of bits of additional information needed to specify it, with all of the microstates treated as equally probable (Gell-Mann 1995). In other words, once we accept the identification of the Clausius thermodynamic entropy with the Boltzmann statistical mechanics entropy, then the interpretation of the entropy as MI is inevitable.

Therefore we are ready to use Jayne's MaxEnt principle to derive the Boltzmann distribution.

As we have seen in the previous chapter, MaxEnt is a general inference principle which was first formulated by E T Jaynes (1957a, 1957b) as:

To describe some partly specified system one should prefer the probability distribution which is compatible with the known information (the system's constraints) and that maximizes the Shannon information entropy.

Jaynes (1957a) summarized his contribution very clearly: '...we accept the Shannon expression for entropy, very literally, as a measure of the amount of uncertainty represented by a probability distribution; thus **entropy becomes the primitive concept with which we work, more fundamental even than energy**. If in addition we reinterpret the prediction problem of statistical mechanics in the subjective sense, we can derive the usual relations in a very elementary way without any consideration of ensembles or appeal to the usual arguments concerning ergodicity or equal *a priori* probabilities. The principles and mathematical methods of statistical mechanics are seen to be of much more general applicability than conventional arguments would lead one to suppose. In the problem of prediction, the maximization of entropy is not an application of a law of physics, but merely a method of reasoning which ensures that no unconscious arbitrary assumptions have been introduced.'

Nevertheless, the maximum-entropy distribution has the important property that no possibility is ignored; it assigns positive weight to every situation that is not absolutely excluded by the given information. And this is quite similar in effect to an ergodic property (Jaynes 1957a).

The basis of MaxEnt is a natural correspondence between statistical mechanics and Shannon's *information theory*. In particular, Jaynes argued that **the entropy of statistical mechanics and the *information entropy* of information theory** (Shannon 1948) **are basically the same thing**. Consequently, statistical mechanics should be seen just as a particular application of a general tool of logical inference and information theory. Interestingly this idea was anticipated by Gibbs (1902): 'But although, as a matter of history, statistical mechanics owes its origin to investigations in thermodynamics, it seems eminently worthy of an independent development, both on account of the elegance and simplicity of its principles, and because it yields new results and places old truths in a new light in departments quite outside of thermodynamics.'

2.2.2 Inferring the canonical probability distribution by the MaxEnt recipe

In this subsection we will closely follow Jaynes (1957a) to derive statistical mechanics as an exercise of entropy maximization and inference. Thus, the idea is to infer the canonical probability distribution, equation (2.26), through the MaxEnt recipe (developed in section 1.4 of the previous chapter). This recipe consists in maximizing the entropy $H = -\sum_{i=1}^{S} P_i \ln P_i$, (we use H to stress that it is the Shannon information entropy, although it coincides with the thermodynamic entropy S except for the multiplicative Boltzmann constant k_B) subject to the known constraints. First, the normalization condition:

$$\sum_{i=1} P_i = 1, \tag{2.44}$$

and second, the given value of the average energy $\overline{E}$ defined as $\sum_{i=1} P_i E_i$, i.e.,

$$\sum_{i=1} P_i E_i = \overline{E}, \tag{2.45}$$

which is equated to the internal energy U of the system. This is done straightforwardly by solving the equation

$$\delta\left[H - \lambda_0\left(\sum_{i=1} P_i - 1\right) - \lambda_1\left(\sum_{i=1} P_i E_i - U\right)\right] = 0, \tag{2.46}$$

where the variation is with respect to each P_i. As before, the Lagrange multiplier λ_0 assures the normalization of the P_i's and λ_1 enforces the known value of the average energy (which is identical to fixing the temperature in the canonical ensemble). The solution is equation (2.26).

Therefore, MaxEnt actually leads to everything we already know since the Boltzmann–Gibbs approach to statistical mechanics. However, MaxEnt is a more general principle which allows us to push the boundaries of statistical physics in new directions and covers a variety of different fields. In fact we realize that MaxEnt works whenever we have a random experiment which has n possible results at each trial; thus in N trials there are n^N conceivable outcomes. (Following Jaynes (1982) we use the word 'result' for a single trial, while 'outcome' refers to the experiment as a whole; thus one outcome consists of an enumeration of N results, including their order.) Each outcome yields a set of sample numbers $\{N_i\}$ and frequencies $\{f_i = N_i/N, 1 \leqslant i \leqslant n\}$, with an information entropy

$$H = -\sum_{i=1}^{n} f_i \ln f_i, \tag{2.47}$$

We can thus conclude that statistical mechanics is just a particular instance of the above setting, in which a system contains N molecules, N_i of which are in the ith state. But it also encloses many different situations, like:

- **Generalized loaded dice**: This is a generalization of the loaded die we analyzed in chapter 1. That is, a die with n faces is tossed N times, the ith face turning up N_i times.
- **Communication**: We receive a message of N symbols, chosen from an alphabet of n letters, the ith letter *occurring* N_i times. We discussed this example in chapter 1.
- **Time series**: Imagine N realizations of a time series $Y \equiv \{y_0, y_1, \ldots, y_T\}$, of which n different sequences $\{Y^{(1)}, Y^{(2)}, \ldots, Y^{(n)}\}$ are possible. The ith sequence $Y^{(i)} = \{y_0^{(i)}, y_1^{(i)}, \ldots, y_T^{(i)}\}$ is realized N_i times. We will use the MaxEnt principle to estimate parameters required for time series forecasting in the last couple of chapters devoted to applications in finance and ecology, respectively.

2.3 Converting information into energy: from Maxwell's demon to Landauer's eraser[3]

In the previous two sections we provided bridges between physics and biology and economics. In section 2.1 we did this at the tools-level, for example the multi-purpose Ising model. In section 2.2 we discussed a conceptual unifying framework, MaxEnt, allowing a synthesis of Shannon's information theory and thermodynamics.

In this section we review another interesting direct relationship between information and thermodynamic entropies; the so-called *Maxwell's demon*. Maxwell's demon is a thought experiment created by the Scottish physicist James Clerk Maxwell. In a letter he wrote in 1867 to his friend, Peter Guthrie Tait, Maxwell imagined that a tiny being who could perceive individual molecules of gas as they move around in a box with a small door between two compartments might be able

[3] The title as well as part of the content this section is based on the article by Lutz and Ciliberto (2015).

violate the Second Law of Thermodynamics (Maxwell 1871):'... *if we conceive of a being whose faculties are so sharpened that he can follow every molecule in its course, such a being, whose attributes are as essentially finite as our own, would be able to do what is impossible to us. For we have seen that molecules in a vessel full of air at uniform temperature are moving with velocities by no means uniform, though the mean velocity of any great number of them, arbitrarily selected, is almost exactly uniform. Now let us suppose that such a vessel is divided into two portions, A and B, by a division in which there is a small hole, and that a being, who can see the individual molecules, opens and closes this hole, so as to allow only the swifter molecules to pass from A to B, and only the slower molecules to pass from B to A. He will thus, without expenditure of work, raise the temperature of B and lower that of A, in contradiction to the second law of thermodynamics.'*

Various forms of Maxwell's 'demon', as William Thomson (Lord Kelvin) later baptized this diminutive being, have been described in the last 150 years. Here we will consider a simpler demon which is more plausible than Maxwell's original one, since it would not need to be able to see or think. That is, an apparatus that creates a pressure difference (rather than a temperature difference), by quickly opening a door between two compartments whenever a molecule, fast or slow, approached from the right, and quickly closing it whenever a molecule approached from the left, would eventually concentrate all the gas on the left (figure 2.5). However, as we have seen before, this would imply a reduction of the gas' entropy because the number of accessible states for each molecule has decreased. This means that, assuming the trapdoor is frictionless, the demon is able to decrease the entropy of the system without producing a corresponding entropy increase elsewhere in the Universe, in apparent violation of the Second Law of Thermodynamics. Or, equivalently, such pressure differential between two compartments can be used to do work, so if one could be established without expending any energy it could form the basis for a

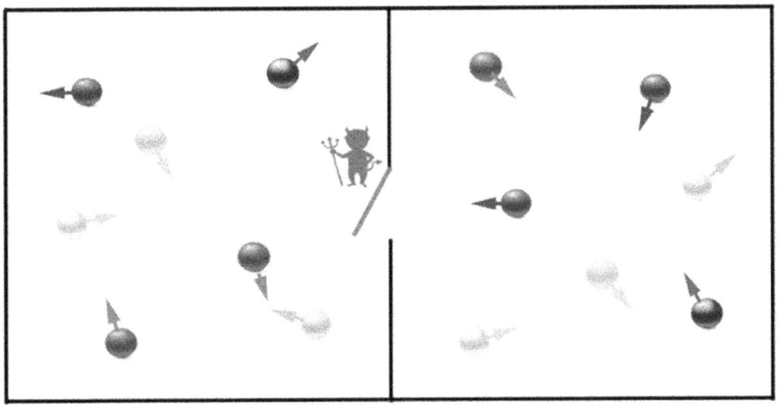

Figure 2.5. Maxwell's demon. Reprinted from Franson (2019), copyright 2019, The Author, with permission of Springer.

perpetual motion device which does work without requiring fuel. It is not immediately evident why such a demon—a one-way valve for molecules—could not be realized as some simple inanimate device, for instance a miniature spring-loaded trapdoor (Bennett 1987) or Feynman's Brownian ratchet device (Feynman *et al* 1963). Nevertheless, as we will see below, no matter how you design your 'demon' component, any device that is able to process and act upon information has an inherent energy requirement that always saves the Second Law.

This seeming paradox caused great controversy among physicists. Indeed, the Second Law of Thermodynamics is one of the most well-established principle in physics. As Eddington (1935) wrote 'If someone points out to you that your pet theory of the universe is in disagreement with Maxwell's equations—then so much the worse for Maxwell's equations. If it is found to be contradicted by observation—well, these experimentalists do bungle things sometimes. But if your theory is found to be against the Second Law of Thermodynamics I can give you no hope; there is nothing for it to collapse in deepest humiliation.' Thus, many physicists have tried to find the flaw in Maxwell's argument. Somehow, somewhere, in the process of looking for molecules of a given type and letting them through the flap, there had to be some entropy generated (Feynman 1996).

For a long time, the prevailing wisdom was that the transfer of information from an object to be measured requires energy dissipation. That is to say, it was generally accepted that entropy was generated as a result of the demon's measurement of the position of the molecules. In fact, one way to *exorcise the demon*, proposed by Brillouin (1962), was to equip him with an electric torch so that he can detect the moving molecules. But such a process would involve dispersing at least one photon, which would cost energy and thus an increase in entropy. This initial increase in entropy is larger than the final decrease, and the overall balance satisfies the Second Law.

However, it was later realized that the dissipation required to save the Second Law and to prevent us from making molecules in thermal equilibrium do work comes not from the measurement process but has a different origin connected with information storing, more precisely with the erasure of information. To understand this it is useful to distinguish two complementary concepts: *information gain* (the reduction in entropy or surprise) and *information erasure*, as *logically reversible* and *irreversible operations* as formulated by Rolf Landauer.

Landauer (1961) investigated the question of what are the physical limitations on building a device to implement a computation. He argued that, since information processing must be carried out by a certain physical system; there should be a one-to-one correspondence between logical states, in terms of zero and one, and physical states. Landauer then distinguished between logically reversible and logically irreversible operations: an operation is logically reversible if the input state can be uniquely identified from the output state. As logically reversible operations need to be one-to-one maps, they correspond to reversible physical processes which, as we have seen, are *isentropic*, i.e., constant entropy processes. And thus they can be implemented by physical devices which do not compress the physical state space. Logically irreversible operations, on the other hand, are noninjective, i.e., many-to-one, mappings. Such operations do not have a unique inverse as there may be many possible original states for a single resulting state. That is, they reduce the logical state space, so must compress the physical state space. Landauer argued that this must be accompanied by a corresponding entropy increase in the environment, in the form of heat dissipation. Another way to see this is that the one-to-one mapping conserves volume in phase space, a result known as *Liouville's Theorem*. According to this theorem, the motion of phase-space representative points (q_i, p_i) conserves volume in phase space. In contrast, dissipative systems usually give rise to flows of phase-space points which contract volumes in phase space. This is because dissipation is expected to reduce the number of accessible asymptotic states, e.g., only the rest state for a real free pendulum with viscous friction.

Suppose we have a two-state system, and let us assume that the two states are occupied with equal probability 1/2, so that the system initially stores the information $H = -1/2 \times \ln(1/2) - 1/2 \times \ln(1/2) = 2 \times 1/2 \times \ln(2) = \ln 2$ (or, if using base 2 for the logarithm, $\log_2 2$ = one bit). Gaining, or writing, information is equivalent to copying information from one place to another, for example mapping the system's zero and one states to the corresponding states of a storage device. Such one-to-one mapping is thus a logically reversible operation that can be realized, in principle, without dissipating any heat. By contrast, the most basic logically irreversible operation consists in erasing a bit of information. This erasure operation takes two input logical states, (conventionally zero and one) and always outputs logical state zero. The state resulting after erasure is commonly denoted as a *standard state*. In terms of physical states, it involves mapping two states (left and right) onto one (right), which is then occupied with probability one. This process reduces the degrees of freedom of the system, which implies a decrease in entropy. In order for this process not to violate the Second Law of Thermodynamics, the energy must be dissipated into the environment. Therefore erasing information in memory entails entropy increase (in the environment). Specifically, Landauer demonstrated that the erasure of one bit of information is necessarily accompanied by the release of at least $k_B T \ln 2$ of heat into the environment. That theoretical result, known as *Landauer's erasure principle*, illustrates a fundamental difference between the process of writing and erasing information.

For a physical implementation of the erasure operation let us consider the so-called Szilard's engine (Szilard 1929), a Maxwell's demon-like apparatus consisting of a single molecule in a box surrounded by a heat bath, that holds the temperature constant, and a partition which divides the box in two compartments (as shown in figure 2.6).

(a) If the molecule is on the left hand side, then the physical state represents logical state '0' and if the molecule is on the right hand side, it represents logical state '1' (figure 2.6(a)).
(b) The partition dividing the box at the center is then removed (figure 2.6(b)).
(c) A piston is inserted at the right end of the box and is slowly moved towards the center of the box (figure 2.6(c)).

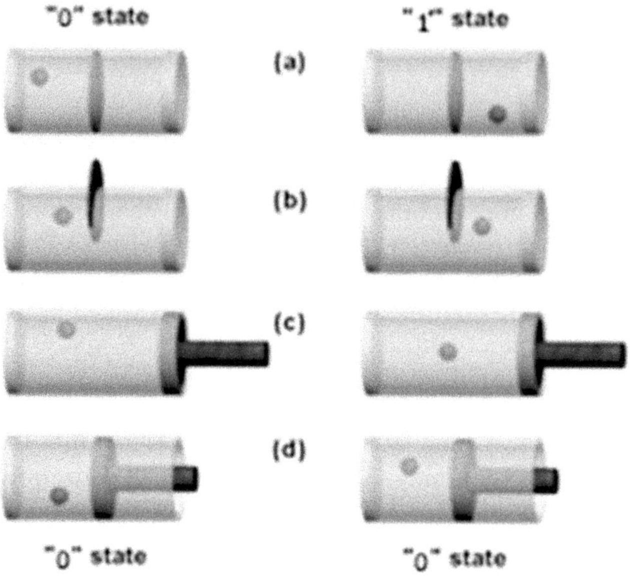

Figure 2.6. Erasure of a logical sate by means of Szilard's engine. Reprinted figure with permission from Maruyama *et al* (2009), copyright (2009) by the American Physical Society.

(d) When the piston reaches the center of the box, the compressed volume becomes $V/2$, the partition is inserted back (figure 2.6(d)).

The resulting standard state is 0 for both initial states, 0 or 1, and the initial information is erased.

The work invested to compress the volume from V to $V/2$ can be calculated in the same way as we did in the example of the isothermal expansion of section 2.1.2:

$$W_{\text{erasure}} = -\int_{V}^{V/2} P\,\mathrm{d}V = \int_{V/2}^{V} \frac{k_B T}{V}\mathrm{d}V = k_B T \ln \frac{V}{V/2} = k_B T \ln 2. \tag{2.48}$$

Since the temperature is kept constant, so is the energy of the molecule and therefore W_{erasure} must be dissipated as heat into the environment, increasing its entropy by $k_B \ln 2$, as Landauer stated.

Landauer's erasure principle was central to solving the paradox of Maxwell's demon. In 1982 Charles Bennett showed that measurements can be carried out reversibly, i.e., without any change in entropy, provided the measuring apparatus is initially in a standard state, so that recording information in the memory does not involve the erasure of information previously stored in the same memory (Bennett 1982). In other words, Maxwell's demon can actually make its measurements **with zero energy expenditure**. And Bennett argued that, according to the erasure principle, the cost comes indeed in the next step, which is the *erasure* of the state 0 or 1 to reset the demon in the state in preparation for the next measurement. After a full cycle of information gathering and energy production, the demon's memory has to be reset to its initial state to allow for a new iteration and the erasure process will always dissipate more energy than the demon produces during one cycle, in full agreement with the Second Law of Thermodynamics.

It turns out that Nature uses molecular-sized motors and machines, which resemble and seem to have the same properties as Maxwell's demon, in virtually every important biological process (Loewenstein 1999, Serreli *et al* 2007, Davies 2019, 2020). Yet, all such 'real demons' comply with the Second Law of Thermodynamics, i.e., they have their entropy-lowering effects balanced by increase of entropy elsewhere. There are plenty of them in our body (Hoffman 2012). Let us mention just a couple of examples of such demon-like systems pervading life. The first one is provided by ultrasensitive switches in molecular biology, like the flagellar motor of *Escherichia coli* bacteria, which switches between clockwise and counterclockwise rotation depending on the intracellular concentration of a regulator protein (Lutz and Ciliberto 2015, Berg 2000). This flagellar motor can be modeled as a Maxwell's demon to calculate the rate of energy consumption needed for both protein sensing and switching and thus providing a quantitative description of the switching statistics (Tu 2008). A second interesting example is at work in our brain. It turns out that neurons can store useful energy, just as Maxwell's demon does by recording distinctions about the positions and velocities of particles, by recording distinctions about the physical states of ions on either side of its trapdoor-like semi-permeable membrane (Aur and Jog 2007, Davies 2019). By the same token, a neuron must expend energy by re-establishing membrane potentials after an action potential has been sent.

Molecular-sized mechanisms are no longer restricted to biology. In fact, the advances in nanotechnology now permit experimental realizations of the Maxwell's demon thought experiment. For instance, David Leigh and coworkers (Serreli *et al* 2007) at the University of Edinburgh, announced the creation of a nano-device based on the Brownian ratchet popularized by Richard Feynman. This synthetic molecular machine can operate by an information ratchet mechanism, in which using light energy, the molecule is able to transmit information about the position of a molecular fragment in a manner that allows transport of the same fragment in a particular direction. So it performs the sorting task envisaged for Maxwell's pressure demon but, crucially, it requires an input of external energy to do so and therefore does not challenge the Second Law of Thermodynamics. Mark Raizen (2009) and coworkers provided another implementation of the process Maxwell envisioned which is capable of sorting individual atoms in a gas into different containers based on their energy. To do this, they first confined atoms in a magnetic trap, which initially are all in the same internal state. Then they introduced a one-way optical barrier, composed of two laser beams arranged side by side: one beam promotes atoms to an excited state, and the other is tuned such that it has no effect on excited atoms but repels atoms in the ground state. This sort of one-way wall for atoms allows reducing the entropy of an ensemble of atoms.

More recently, in 2014, Koski *et al* created an experimental realization of a Szilárd engine in which the role of the gas molecule is played by a single electron confined to a two-sided nanoscale box that is coupled to a heat bath (Koski *et al* 2014). Furthermore, so far, all realizations of a Maxwell's demon had required an external control of its functions, making it hard to quantitatively evaluate the thermodynamic parameters—heat, entropy and information transfer—involved in the process. This is an autonomous Maxwell's demon that works without external intervention. What makes the demon autonomous or self-contained is that it performs the measurement and feedback operation without outside help. In fact, the role of the demon is played by another single-electron box coupled to the first. The primary single-electron box was maintained at the dilution-refrigerator temperatures in the 0.1 K range. Such extremely low temperatures are required in order for the system to be so well isolated that it is possible to register extremely small temperature changes. This system performs a closed cycle and thus energy conservation makes the total heat Q extracted from the reservoir equal to the work W extracted from the engine. The cycle exploits the existence of two degenerate box states for a certain electron energy. The cycle begins with the electron in a definite, nondegenerate state. The electron is thermally excited to the degenerate level, where the electron can reside with equal probability in either of the two states. That introduction of information entropy of one bit represents an increase in the thermodynamic entropy of the electron of $k_B T \ln 2$ and a corresponding decrease in the entropy, in the refrigerator, and thus the temperature, of the bath. Then the other single-electron box (the demon) coupled to the first detects which of the two states the electron is in and autonomously feeds the information to rapidly return the electron to its initial nondegenerate state and complete the cycle. A main finding is that the creation of one bit of information per cycle—which state the electron is in—could extract heat from the bath with an average efficiency of about 75%. So in a way Maxwell was right: information really can serve as a type of fuel.

Therefore, the main lesson of this section is that information is not just an abstract concept but a physical quantity: information can be converted to net energy. In other words, the connection between information and physics can be regarded as a two-way street: on the one hand, we had seen the epistemological perspective, i.e., the fact that all of our physical understanding is rooted in information. On the other hand, we just learned that information is physical, that is, the ontological perspective. Indeed, *Information physics* (Knuth 2010, Dittrich 2015) is a fundamentally new approach to science based on this duality of information-based physics, i.e., physics 'made of' information, as well as physically based information, i.e., information 'made of' physics (Perdigão *et al* 2020).

2.4 Conclusion

Entropy is a central concept of modern science and technology. In this chapter we started with Clausius' original definition of entropy, connected to the **irreversibility** of real thermodynamic processes.

Boltzmann later interpreted entropy as the measure of the number of possible microscopic arrangements or states of individual atoms or molecules of a physical system that comply with the macroscopic condition of the system (e.g., the volume and temperature in the case of a gas). He thereby introduced the concept of statistical **disorder** and a formulation in terms of probability distributions creating a new field of thermodynamics, called statistical mechanics. This allowed Boltzmann to derive the increase in entropy in irreversible processes by combining the laws of mechanics with the laws of probability. That is, he explained the Second Law of Thermodynamics arising simply from statistical considerations.

As we mentioned in chapter 1, from an information theory perspective, entropy is understood as **missing information** (MI) and is a synonym of **ignorance** or **uncertainty**. So, from the point of view of information theory, a message with high Shannon entropy means that its information content is also high. Jaynes subsequently showed that Shannon information entropy and Boltzmann thermodynamic entropy is basically one and the same thing. Figure 2.7 is a schematic representation of the connections between these three different interpretations of entropy.

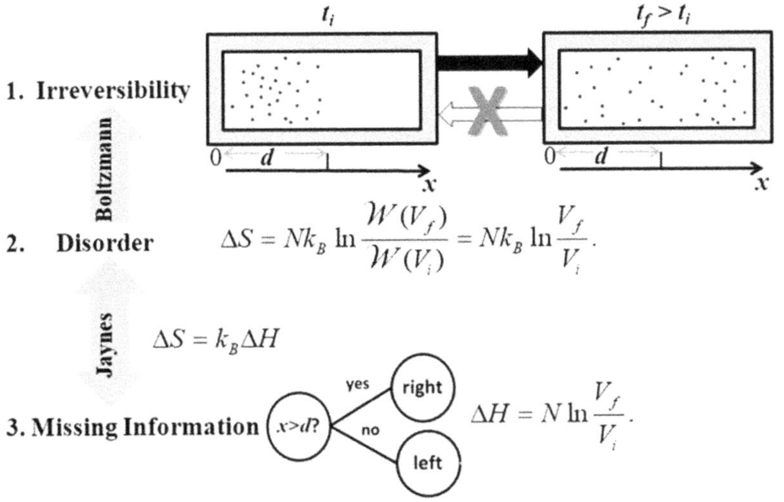

Figure 2.7. Scheme showing how the three definitions of entropy are connected.

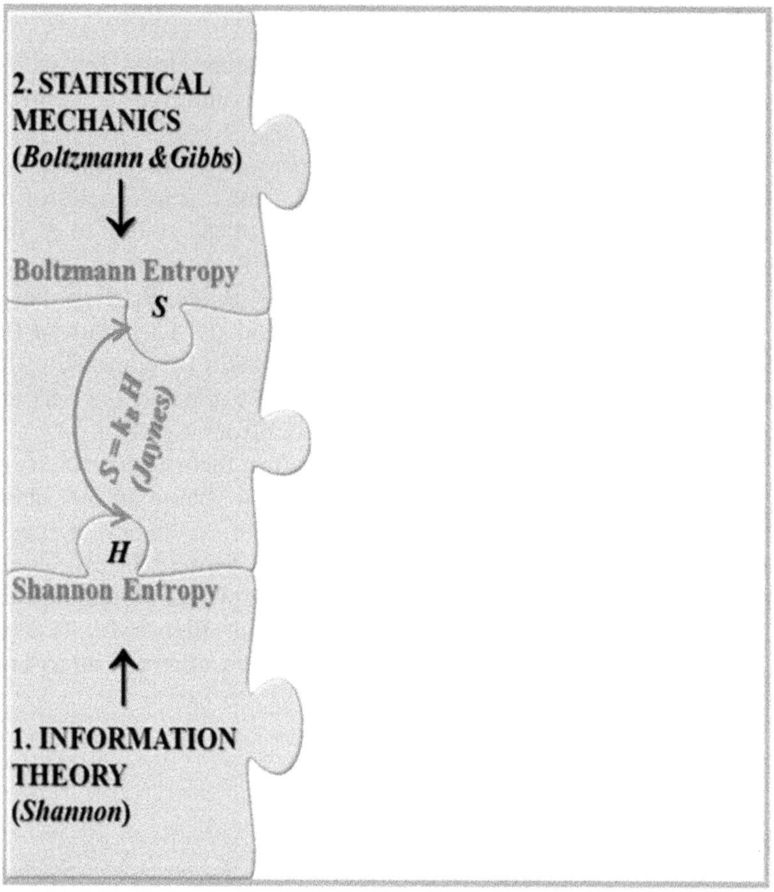

References

Aur D and Jog M 2007 Beyond spike timing theory – thermodynamics of neuronal computation *Nat. Prec.* https://doi.org/10.1038/npre.2007.1254.1

Ausloos M *et al* 1999 Applications of statistical physics to economic and financial topics *Physica A* **274** 229–40

Bai F *et al* 2010 Conformational spread as a mechanism for cooperativity in the bacterial flagellar switch *Science* **327** 685–9

Bennett C H 1987 Demons, engines, and the second law *Sci. Am.* **257** 108–16

Bennett C H 1982 The thermodynamics of computation—a review *Int. J. Theor. Phys.* **21** 905–40

Berg H C 2000 Motile behavior of bacteria *Phys. Today* **53** 24–9

Boltzmann L 1896 *Lectures on Gas Theory* (New York: Dover) (reprint)

Boltzmann L 1877 Über die Beziehung zwischen dem zweiten Hauptsatz der mechanischen Wärmetheorie und der Wahrscheinlichkeitsrechnung respektive den Sätzen über des Wärmegleichgewicht' *Wiener Berichte* **76** 373–435

Brillouin L N 1962 *Science and Information Theory* 2nd edn (New York: Academic)

Brock W A and Durlauf S N 2001 Discrete choice with social interactions *Rev. Econ. Stud.* **68** 235–60
Callen H B 1985 *Thermodynamics and an Introduction to Thermostatistics* (Hoboken, NJ: Wiley)
Carnot S 1824 Réflexions sur la puissance motrice du feu et sur les machines propres à développer cette puissance (Paris: Bachelier) (in French)
Castellano C, Fortunato S and Loreto V 2009 Statistical physics of social dynamics *Rev. Mod. Phys.* **81** 591
Clausius R 1865 Ueber verschiedene für die Anwendung bequeme Formen der Hauptgleichungen der mechanischen Wärmetheorie *Ann. Phys.* **125** 353–400
C G K 1907 The scientific papers of J Willard Gibbs *Nature* **75** 361–2
Davies P 2020 Does new physics lurk inside living matter? *Phys. Today* **73** 34–40
Davies P 2019 *The Demon in the Machine: How Hidden Webs of Information are Solving the Mystery of Life* (Chicago, IL: University of Chicago Press)
Dittrich T 2015 The concept of information in physics': an interdisciplinary topical lecture *Eur. J. Phys.* **36** 015010
Eddington A 1935 *New Pathways in Science* (Cambridge: Cambridge University Press)
Feynman R P 1996 *Feynman Lectures on Computation* (Reading, MA: Addison-Wesley)
Feynman R P, Leighton R B and Sands M 1963 *Feynman Lectures on Physics* vol I (Reading, MA: Addison-Wesley)
Fort H 2013 Statistical mechanics ideas and techniques applied to selected problems in ecology *Entropy* **15** 5237–76
Franson J D 2019 Velocity-dependent optical forces and Maxwell's demon *Sci. Rep.* **9** 13798
Glazer M and Wark J 2006 Statistical Mechanics *A Survival Guide* (New York: Oxford University Press)
Gibbs J W 1902 *Elementary Principles in Statistical Mechanics* (New Haven, CT: Yale University Press)
Gell-Mann M 1995 *The Quark and the Jaguar* 3rd edn (London: St. Martin's Press)
Harte J 2011 *Maximum Entropy and Ecology: A Theory of Abundance, Distribution, and Energetics* (Oxford: Oxford University Press)
Hoffman P 2012 *Life's Ratchet: How Molecular Machines Extract Order from Chaos* (New York: Basic Books)
Hohnisch M, Pittnauer S, Solomon S and Stauffer D 2005 Socioeconomic interaction and swings in business confidence indicators *Physica* A **345** 646–56
Hopfield J J 1982 Neural networks and physical systems with emergent collective computational abilities *Proc. Natl Acad. Sci.* **79** 2554–58
Ising E 1925 Beitrag zur Theorie des Ferromagnetismus *Z. Phys.* **31** 253–58
Jaynes E T 1982 On the rationale of maximum-entropy methods *Proc. IEEE* **70** 939–53
Jaynes E T 1978 Where do we stand on Entropy? *Presented at Maximum Entropy Formalism Conf. (May 2–4, 1978)*
Jaynes E T 1957a Information theory and statistical mechanics I *Phys. Rev* **106** 620–30
Jaynes E T 1957b Information theory and statistical mechanics II *Phys. Rev.* **108** 171–90
Knuth K H 2010 Information physics: the new frontier *AIP Conf. Proc.* **1305** 3–19
Koski J V, Maisi V F, Pekola J P and Averin D V 2014 Experimental realization of a Szilard engine with a single electron *Proc. Natl Acad. Sci.* **111** 13786–9
Landauer R 1961 Irreversibility and heat generation in the computing process *IBM J. Res. Dev.* **5** 183–91

Landau L and Lifshitz E M 1980 *Statistical Physics. Third Revised and Enlarged Edition* (Oxford: Pergamon)

Lenz W 1920 Beiträge zum Verständnis der magnetischen Eigenschaften in festen Körpern *Phys. Z.* **21** 613–5

Loewenstein W 1999 *The Touchstone of Life: Molecular Information, Cell Communication, and the Foundations of Life* (New York: Oxford University Press) p 227

Lutz E and Ciliberto S 2015 Information: from Maxwell's demon to Landauer's eraser *Phys. Today* **68** 30–5

Lux T 2007 Applications of statistical physics in finance and economics *Economics Working Papers 2007–05* (Kiel: Department of Economics, Christian-Albrechts-University of Kiel)

Maruyama K, Nori F and Vedral V 2009 Colloquium: the physics of Maxwell's demon and information *Rev. Mod. Phys.* **81** 1–23

Maxwell J C 1867 On the dynamical theory of gases *Phil. Trans. R. Soc. Lond.* **157** 49–88

Maxwell J C 1871 *Theory of Heat. Reprinted 2011* (Cambridge: Cambridge University Press)

Maynard-Smith J 1978 *Models in Ecology* (Cambridge: Cambridge University Press)

Moore C C 2015 Ergodic theorem, ergodic theory, and statistical mechanics *Proc. Natl Acad. Sci.* **112** 1907–12

Onsager L 1944 Crystal statistics. I. A two-dimensional model with an order–disorder transition *Phys. Rev.* **65** 117–49

Pathria R K and Beale P D 2011 *Statistical Mechanics* 3rd edn (New York: Elsevier)

Perdigão R A P, Ehret U, Knuth K H and Wang J 2020 Debates: does information theory provide a new paradigm for Earth science? Emerging concepts and pathways of information physics *Water Resour. Res.* **56** e2019WR025270

Raizen M G 2009 Comprehensive control of atomic motion *Science* **324** 1403–6

Serreli V, Lee C F, Kay E U and Leigh D A 2007 A molecular information ratchet *Nature* **445** 523–7

Shannon C E 1948 A mathematical theory of communication *Bell Syst. Tech. J.* **27** 379–423

Shi Y and Duke T 1998 Cooperative model of bacterial sensing *Phys. Rev.* E **58** 6399–406

Sornette D 2014 Physics and financial economics (1776–2014): puzzles, Ising and agent-based models *Rep. Prog. Phys.* **77** 062001

Stauffer D 2008 Social applications of two-dimensional Ising models *Am. J. Phys.* **76** 470–3

Szilard L 1929 On the decrease of entropy in a thermodynamic system by the intervention of intelligent beings *Z. Angew. Phys.* **53** 840–56

Feld B T and Weiss Szilard G (ed) 1972 *English translation in The Collected Works of Leo Szilard: Scientific Papers* (Cambridge, MA: MIT Press) pp 103–29

Tu Y 2008 The nonequilibrium mechanism for ultrasensitivity in a biological switch: sensing by Maxwell's demons *Proc. Natl Acad. Sci. USA* **105** 11737

Voit J 2005 *The Statistical Mechanics of Financial Markets* (Berlin: Springer)

Vtyurina N N *et al* 2016 Hysteresis in DNA compaction by Dps is described by an Ising model *Proc. Natl Acad. Sci.* **113** 4982–7

Chapter 3

Elements of physical biology: the Lotka–Volterra equations

'It would seem, then, that what is needed is an altogether new instrument; one that shall envisage the units of a biological population as the established statistical mechanics envisage molecules, atoms and electrons; that shall deal with such average effects as population density, population pressure, and the like, after the manner in which thermodynamics deal with the average effects of gas concentration, gas pressures, etc; that shall accept its problems in terms of common biological data, as thermodynamics accepts problems stated in terms of physical data; and that shall give the answer to the problem in the terms in which it was presented. What is needed, in brief, is something of the nature of what has been termed <<Allgemeine Zustandslehre>>, a general method or Theory of State. It is somewhat along these lines that the system now to be sketched is conceived.'

—A J Lotka *Elements of Physical Biology* (1925)

- We start this chapter with the simplest growth equation for a single isolated population, the Malthus' equation (1798). The problem with this equation is that it assumes a constant per capita growth rate, leading to exponential growth. This drawback was overcome by Verhulst in 1838 assuming resource limitation, which means replacing the constant per capita growth rate by a ***density dependent*** per-capita growth rate, the resulting ***logistic equation***, saturates the population to an equilibrium value called the ***carrying capacity***.
- However, in the real world, species are rarely isolated from other species. So, to model the effect of the interspecific interactions on the dynamics of each species, we introduce the linear Lotka–Volterra generalized equations (LLVGE). We show that, by estimating the LLVGE parameters from the yields in monoculture and biculture experiments, the LLVGE produce quite accurate predictions for species yields in single-trophic level communities of $S > 2$ species, either artificial or natural.

doi:10.1088/978-0-7503-3931-5ch3

- It turns out that the LLVGE are similar to the equations used in chemical kinetics. Indeed, when proposing these LLVGE, Lotka sought to develop for biological systems involving different interacting species a statistical mechanics approach similar to the one Boltzmann and Gibbs had developed to deal with the kinetic theory of gases and related problems. Therefore, in section 3.3 we start by briefly discussing different attempts to approach ecosystems from statistical mechanics. In particular, cellular automata (CA) constitute a simple statistical mechanics modelling tool to investigate the spatio-temporal dynamics of several complex systems. We thus introduce two cellular automata to model ecological problems, namely desertification of semi-arid lands by overgrazing and the evolution of communities of trees in tropical forests.

3.1 The kinetic formulation of population dynamics

3.1.1 One isolated species

The mathematical theory of population dynamics originated in attempts to describe the growth of human or animal populations. The most elementary form is the exponential or Malthus' law. Imagine we have a population of $n(t)$ individuals at time t and we want to know how it will change with time. Population changes are the result of inputs, births and immigration, minus outputs, deaths and emigration. Let us assume the simplest situation without migration in which the above relation can be expressed mathematically as:

$$\frac{\mathrm{d}n(t)}{\mathrm{d}t} = (b - d)n(t), \tag{3.1}$$

where b and d are ***constant*** per-capita rates, measured, respectively, as births and deaths per individuals per unit of time. It is customary to group $(b - d)$ into a net constant per-capita growth rate r and write equation (3.1) as:

$$\frac{1}{n(t)}\frac{\mathrm{d}n(t)}{\mathrm{d}t} = r. \tag{3.2}$$

We know from calculus that the solution of this equation is

$$n(t) = n_0 \mathrm{e}^{rt}, \tag{3.3}$$

where $n_0 = n(0)$ is the initial population. Thus if $r > 0$ (i.e., $b > d$) the population grows exponentially while if $r > 0$ (i.e., $b < d$) it dies out, as shown in figure 3.1(a).

In fact, the differential equation (3.3), due to Malthus in 1798, is fairly unrealistic. It says that the rate of change in population size over time, $\mathrm{d}n(t)/\mathrm{d}t$, is proportional to the current population size, $n(t)$. This is an example of a density-independent biological force, because it depends only on the population, n, not on external forces such as crowding or food supply. In other words, in deriving equation (3.1) it was unrealistically assumed that per-capita rates b and d were constant, independent from the population density, and thus the same happens for $r = b - d$ in equation (3.2). However, when resources (nutrients, water, space, etc) are limited, if crowding is

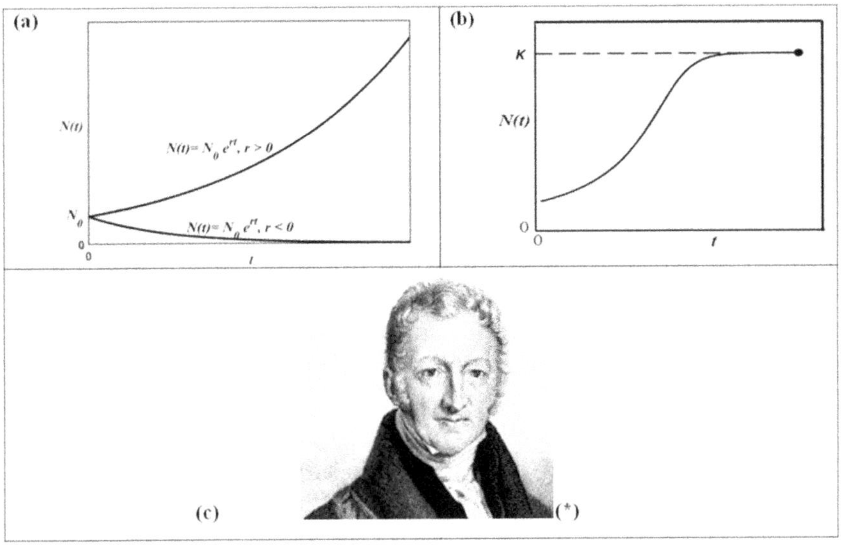

Figure 3.1. (a) Malthus equation (3.2): exponential growth ($r > 0$) and decay ($r < 0$). (b) Verhulst's equation: logistic growth. (c) Thomas Robert Malthus (1766–1834) (this Thomas Malthus image has been obtained by the author from the Wikimedia website, where it is stated to have been released into the public domain. It is included within this article on that basis.

increased, one expects the per capita birth rate to decrease and the per capita death rate to increase because fewer resources are available for organisms to use for reproduction and survival, respectively. Hence, these per-capita rates should really be a function of the population density of the species, so that, as population density increases to sufficiently high levels, the death rate rises and the birth rate falls, bringing about a fall in r toward zero and below, into negative values. Thus, more realistic per-capita rates b and d should exhibit *density dependence*.

The rough law was improved in 1838 by the Dutch mathematician and biologist Pierre Verhulst, who proposed the simplest formula for a decreasing net growth rate $r(n)$ is a linear one, i.e., straight line with a proportionality constant c:

$$r(n) \equiv r - cn = r - cn = r\left(1 - \frac{n}{r/c}\right). \tag{3.4}$$

If we define $K \equiv r/c$ and replace r in (3.2) by (3.4), we arrive at Verhulst's *logistic equation*:

$$\frac{1}{n(t)}\frac{\mathrm{d}n(t)}{\mathrm{d}t} = r\left(1 - \frac{n(t)}{K}\right), \tag{3.5}$$

where K is interpreted as the so-called *carrying capacity*, i.e., a maximum sustainable population attained at equilibrium, since $\mathrm{d}n/\mathrm{d}t = 0$ for $n = K$, as shown in figure 3.1(b). Therefore, the per capita population growth rate $(1/n)(\mathrm{d}n/\mathrm{d}t)$ is at its maximum value of r when n is close to zero, then declines linearly to zero when n reaches K. If n exceeds K, the per capita growth rate becomes negative.

3.1.2 Many interacting species

To model the dynamics of a community of interacting populations of several species it is necessary to derive a mathematical representation for the process of two individuals, each of them of an arbitrary species, to meet each other in order to interact. It turns out that ecology, being an interdisciplinary subject, has often advanced by borrowing perspectives and methods from numerous other disciplines, like mathematics, physics and chemistry. Inspired in such disciplines, two fundamental works giving rise to the systematic study of population dynamics were published in 1925 and 1926. They tackled the problem of the interaction between different species. These works were the *Elements of Physical Biology* (1925) by the American physical chemist A J Lotka, and a paper by the Italian physicist and mathematician Vito Volterra (1926).

The goal of Lotka with his book was to create a new biological discipline called 'physical biology,' by which he meant the 'broad application of physical principles and methods in the contemplation of biological systems'. The source of inspiration for him was chemical kinetics, also known as reaction kinetics, the branch of physical chemistry that is concerned with understanding the rates of chemical reactions. Therefore, when two compounds A and B in a dilute gas or a solution react to yield a product C, the rate at which this chemical reaction takes place is assumed to be proportional to the densities of both reactants A and B. Hence, for the chemical reaction A + B → C the rate with which the product C is produced is proportional to their concentrations [A]·[B]. The analogy between physical chemistry and physical biology lay in the idea that the same basic laws governed both kinds of systems. In the chemical system the components were molecules while in the biological system the components were organisms plus the raw materials in their environment, and the exchanges of matter and energy took place through the web of food relationships, growth, and reproduction. That is, **Lotka sought to develop for such a biological system a statistical mechanics approach similar to the one Boltzmann and Gibbs had developed to deal with the kinetic theory of gases and related problems** we discussed in chapter 2.

Using this analogy with chemical kinetics, Lotka proposed for S interacting species a set of equations given by:

$$\frac{\mathrm{d}n_i(t)}{\mathrm{d}t} = r_i\left(n_i(t) + \sum_{j=1}^{S} I_{ij} n_i(t) n_j(t)\right) i = 1, \ldots, S, \text{ which can be re-written as}$$

$$\frac{1}{n_i(t)}\frac{\mathrm{d}n_i(t)}{\mathrm{d}t} = r_i\left(1 + \sum_{j=1}^{S} I_{ij} n_j(t)\right) \qquad i = 1, \ldots, S, \tag{3.6}$$

where r_i is the intrinsic growth rate of species i, with dimension of time^{-1}, and I_{ij} is an interaction coefficient quantifying the strength of the effect of species j on species i through their pairwise interaction, which can be either negative, positive or zero, with dimension of population^{-1}. Volterra independently performed the analysis of predator–prey interactions, publishing a short discussion in his 1926 paper. In subsequent papers and books (1931, 1935), he extended this approach to cover interspecific competition and arriving at the same kind of modelling. This is why equation (3.6) is called the *linear* Lotka–Volterra *generalized* equations (LLVGE). Generalized means that I_{ij} can be <0, =0 or >0. While the word 'linear' comes from the fact that these equations can be thought as the first order or linear approximation in a Taylor series expansion of the per-capita growth rates of species about the equilibrium points of a more complex and general theory (Lotka 1925, Volterra 1926, 1931):

$$\frac{1}{n_i(t)}\frac{\mathrm{d}n_i(t)}{\mathrm{d}t} = f_i(n_1, n_2, \ldots, n_S) \qquad i = 1, \ldots, S, \tag{3.7}$$

where $f_i(n_1, n_2, \ldots, n_S)$ are arbitrary functions of the S species making up the community, which we denote more compactly in terms of the S-dimensional vector of yields $\mathbf{n} = [n_1, n_2, \ldots, n_S]$ as $f_i(\mathbf{n})$. Therefore, let $\delta\mathbf{n} = \mathbf{n} - \mathbf{n}^*$ be a small displacement from an equilibrium point $\mathbf{n}^*$, which by definition satisfies $f_i(\mathbf{n}^*) = 0$, and if we perform a Taylor expansion of $f_i(\mathbf{n})$ around $\mathbf{n}^*$ we get:

$$\begin{aligned} f_i(\mathbf{n}) &= f_i(\mathbf{n}^*) + \sum_{j=1}^{S} \frac{\partial f_i(\mathbf{n})}{\partial n_j}\bigg|_{\mathbf{n}*} \delta n_i + \sum_{j=1}^{S}\sum_{k=1}^{S} \frac{\partial^2 f_i(\mathbf{n})}{\partial n_j \partial n_k}\bigg|_{\mathbf{n}*} \delta n_j \delta n_k + \\ &\quad \sum_{j=1}^{S}\sum_{k=1}^{S}\sum_{l=1}^{S} \frac{\partial^3 f_i(\mathbf{n})}{\partial n_j \partial n_k \partial n_l}\bigg|_{\mathbf{n}*} \delta n_j \delta n_k \delta n_l + \cdots \\ &= 0 + \sum_{j=1}^{S} \frac{\partial f_i(\mathbf{n})}{\partial n_j}\bigg|_{\mathbf{n}*} \delta n_j + O(\delta n^2) \\ &= -\sum_{j=1}^{S} \frac{\partial f_i(\mathbf{n})}{\partial n_j}\bigg|_{\mathbf{n}*} n_j^* + \sum_{j=1}^{S} \frac{\partial f_i(\mathbf{n})}{\partial n_j}\bigg|_{\mathbf{n}*} n_j + O(\delta n^2). \end{aligned} \tag{3.8}$$

where $O(\delta N^2)$ involve quadratic terms given by $\sum_{j=1}^{S}\sum_{k=1}^{S} \frac{\partial^2 f_i(\mathbf{n})}{\partial n_j \partial n_k}\Big|_{\mathbf{n}*} \delta n_j \delta n_k$, or terms of order greater than two, i.e., cubic terms given by $\sum_{j=1}^{S}\sum_{k=1}^{S}\sum_{l=1}^{S} \frac{\partial^3 f_i(\mathbf{n})}{\partial n_j \partial n_k \partial n_l}\Big|_{\mathbf{n}*} \delta n_j \delta n_k \delta n_l$.

Hence, we re-obtain equation (3.6) by identifying

$$r_i = -\sum_{j=1}^{S} \frac{\partial f_i(\mathbf{n})}{\partial n_j}\bigg|_{\mathbf{n}*} n_j^*, \tag{3.9a}$$

$$r_i I_{ij} = \frac{\partial f_i(\mathbf{n})}{\partial n_j}\bigg|_{\mathbf{n}*}. \tag{3.9b}$$

Let us introduce at this point a widely used matrix in community ecology, which is closely related with the matrix $[I_{ij}]$, the *community matrix* $[J_{ij}]$ (Levins 1968). The community matrix is nothing but the mathematical Jacobian matrix. That is, in vector calculus, the Jacobian matrix of a vector-valued function in several variables is the matrix of all its first-order partial derivatives:

$$J_{ij} \equiv \left.\frac{\partial(n_i f_i)}{\partial n_j}\right|_{\mathbf{n}*} = \left.\frac{\partial\left(\frac{\mathrm{d}n_i}{\mathrm{d}t}\right)}{\partial n_j}\right|_{\mathbf{n}*}. \tag{3.10}$$

It can be shown that the matrix $[I_{ij}]$ and $[J_{ij}]$ are connected by a simple relationship (see Appendix at the end of this chapter):

$$I_{ij} = \frac{J_{ij}}{r_i n_i *}. \tag{3.11}$$

It is worth remarking that linear Lotka–Volterra equations are regarded as *descriptive* or *phenomenological* models, i.e., they describe how the abundance of one species affects the abundance of another, without specifically including a particular mechanism for such interaction (Morin 2011). Neither the nature of the particular competition mechanism nor if it is direct or mediated by other(s) species, is specified. Rather, *effective* interaction coefficients summarize the per capita effects of one species on another. Depending on the signs of I_{ij} and I_{ji} for the pair of species i and j we have the following possibilities:

- mutual competition –/–, each species has an inhibiting effect on the growth of the other;
- amensalism –/0, the growth of one species is negatively affected while the growth of the other is unaffected;
- predation –/+[1], the 'predator' ('prey'), has an inhibiting (accelerating) effect on the growth of the prey (predator);
- commensalism 0/+, the growth of one species accelerates while the other is unaffected;
- mutual cooperation or mutualism +/+, each species has an accelerating effect on the growth of the other.

A continuous or a discrete time description?
So far we have used a continuous time description, in terms of differential equations. However, it is also possible to use a discrete time formulation in terms of *difference equations*. What should we use, a continuous or discrete time description? In some

[1] Notice that according to our definitions, the host–parasite and the plant–herbivore interactions would be classified as 'predation'.

systems changes take place in discrete time intervals, like the depositing of interest in a bank account, while in other systems, like the column of a mercury thermometer when measuring the body temperature, the change happens continuously. Difference equations better represent the first kind of systems while differential equations are more suitable for the latter. From a biological point of view, if births occur continuously with overlapping generations in relatively aseasonal environments a continuous overlap of generations, the continuous time is a sound choice. However, for many species births occur in regular time intervals so they have no overlap whatsoever between successive generations and therefore population growth is better-suited for a discrete-time description. For primitive organisms such discrete steps can be quite short in which case a continuous time model may be a reasonable approximation. In fact, the step lengths can vary widely from species to species.

On the one hand, difference equations reflect an essential property of the real world, namely its discreteness. They are also appealing due to their simplicity. Indeed, only quite simple computational and graphical representation tools are necessary to study the behavior of the solutions of difference equations and their bifurcations for changing parameters. Other examples of problems more amenable to finite difference treatment are the formation of structures in turbulence, particularly the cascade formation of vortices of successively decreasing scales, or complicated oscillating processes in complex electric circuits (Sharkovsky *et al* 1993). On the other hand, the advantage of using differential equations is that qualitative insight is usually gained from simple model problems that may be solved using analytical methods. That is, as we shall see in a moment, calculus often helps in uncovering functional relationships between the relevant problem variables. However, most problems of interest lead to differential equations that cannot be solved easily using analytic techniques. In such cases numerical methods allow us to use the power of a computer to obtain quantitative insight. Since, a computer is limited to finite combinations of the four arithmetic operations, +, −, ×_, ÷_, and logical operations, numerical methods require discrete time for running such computations. Therefore, we will resort to the continuous time description to introduce the formalism of population dynamics. However, for most practical applications, it seems more natural to build the model as a *discrete* difference equation from the start, without going through the doubly approximative process of first, during the modelling stage, finding a differential equation to approximate a basically discrete situation, and then, for numerical computing purposes, approximating that differential equation by a difference scheme (for an interesting discussion see van der Vaart 1973)[2].

[2] A note of caution is in order here. As was shown by Robert May (1976), some of the simplest nonlinear difference equations can exhibit a wide spectrum of dynamical behavior. From stable equilibrium points, to stable cyclic oscillations between two population points, four points, eight points, etc, through to a chaotic regime in which (depending on the initial population value) cycles of any period, or even totally aperiodic but bounded population fluctuations, can occur.

3.2 The Lotka–Volterra linear model for single-trophic communities

All life forms can be broadly lumped into one of two categories: the autotrophs, who produce organic matter from inorganic substances (like plants, algae) and the heterotrophs, organisms that cannot produce their own food, relying instead on the consumption of other organisms (either autotrophs or heterotrophs).

Indeed, the heterotrophs can be arranged in different levels according to who eats whom, called trophic levels. So, for example, the green plants (producers) are the first trophic level, the organisms that feed on plants are the second level (primary consumers), carnivores are the third level and carnivore predators are the fourth level (secondary and tertiary consumers). This is an example of a general *trophic chain* or *food chain*.

In this section we particularize equation (3.6) for a single-trophic community, e.g., herbivores or plants. We do this because we will use these equations for different applications in subsequent chapters. Thus the trophic level below the community of interest, corresponding to the food required for individuals of each species i to thrive (plants in the case of herbivores; soil nutrients in the case of plants) is modeled through the species carrying capacity K_i. This implies setting

$$I_{ij} = \alpha_{ij}/K_i \, , \tag{3.12}$$

where α_{ij} are interaction coefficients measuring the per capita effect of species j on the abundance of species i.

As we have said, the LLVGE can accommodate all kind of possible pairs of sign interactions. Indeed, in the past single-trophic communities were synonymous with mutually negative competition, –/–, for resources. Interaction between species, like –/+ or +/+, had been considered to occur only between different trophic levels (Morin 2011). For instance, the typical example of –/+ is predation and of +/+ is mutualism, like the plant–pollinator relationship. However, several examples of these types of interspecific interactions have also been found for species sharing the same trophic level; from viruses (Turner and Chao 1999, Arbiza *et al* 2010) to natural plant communities (Holmgren *et al* 1997) or artificial plant polycultures (Halty *et al* 2017), just to mention few examples.

Therefore, by equation (3.12), the LLVGE for S interacting species within a single-trophic level can be written as:

$$\frac{dn_i}{dt} = r_i n_i \left(1 + \frac{\sum_{j=1}^{S} \alpha_{ij} n_j}{K_i} \right) \qquad (\alpha_{ii} = -1) i = 1, \dots, S. \tag{3.13}$$

The choice $\alpha_{ii} = -1$ for all species i in equation (3.13) is to warrant that in the case of just one species it reduces to the logistic equation (3.5). Notice that this means we are 'normalizing' the strength of the interspecific interactions a species receives by its intraspecific interaction strength.

At this point it is worth summarizing the definitions and interpretations of the three alternative interaction strength matrices we have introduced (table 3.1, adapted from Novak *et al* 2016).

Table 3.1. Summary of the definitions and interpretations of the three alternative interaction strength matrices.

	Community matrix	Interaction matrix	Lotka–Volterra generalized matrix
Mathematical definition	$J_{ij} \equiv \left.\dfrac{\partial\left(\dfrac{dn_i}{dt}\right)}{\partial n_j}\right\vert_{n*}$	$I_{ij} \equiv \left.\dfrac{\partial\left(\dfrac{1}{r_i n_i}\dfrac{dn_i}{dt}\right)}{\partial n_j}\right\vert_{n*}$	$\alpha_{ij} \equiv \left.\dfrac{\partial\left(\dfrac{K_i}{r_i n_i}\dfrac{dn_i}{dt}\right)}{\partial n_j}\right\vert_{n}$ *
Defined for	any model of differentiable functions $f_i(n_1, n_2, \ldots, n_S)$.	any model of differentiable functions $f_i(n_1, n_2, \ldots, n_S)$.	the linear generalized Lotka–Volterra model.
Dimensions	time^{-1}	$N^{-1}\ L^2$ or $M^{-1}\ L^2$ (†)	non-dimensional
Interpretation	Direct effect of the average species individual j on species i's population growth rate	per-capita growth rate relative to its intrinsic growth rate	per-capita growth rate relative to its intrinsic growth rate

(*)In general you work with densities, i.e., number of individuals per unit of area, or ***yields*** (biomass per unit of area).

3.2.1 Obtaining the model parameters from monoculture and biculture experiments

The LLVGE constitute the simplest mathematical model for a community of S interacting species. Its minimality of the LLVGE often raises doubts on their ability to make quantitative predictions, like species abundances. Lotka–Volterra models are instead regarded more as a qualitative than a quantitative tool in population or community ecology (Brown *et al* 2001). More 'realistic' and complex theories, e.g., equations involving nonlinear response functions (Vandermeer and Goldberg 2013) and higher order interactions (Abrams 1983) implying non-additive effects (Morin 2011), are often preferred because they are perceived as more reliable (although they can be as intractable as the real systems they aim to model). However, this is done at the price of including additional parameters which are very hard to measure. Actually, in many natural communities—like tropical forests, plankton or mutualistic networks—the species richness S is of the order of hundreds. Hence, even estimating all the parameters of the LLVGE from empirical data is an unfeasible task, let alone estimating additional parameters of more complex models.

Our viewpoint is that models should be mainly evaluated not on the basis of the realism of their assumptions but on the basis of the accuracy of their predictions. So our goal is to analyze how well the LLVGE work as a *quantitative* tool for describing/

explaining/predicting the outcome of experiments for single-trophic species belonging to the same taxonomic group (plants, algae, etc) assuming that a state of equilibrium was reached. Specifically, we want to test the accuracy of LLVGE for predicting the yields (biomass per unit of area), n_i. With this aim we have to obtain the model parameters, $\{r_i\}$, $\{K_i\}$ and $\{\alpha_{ij}\}$. If we focus on the equilibrium yields, $\{n_i^*\}$, rather than on the species time trajectories $\{n_i(t)\}$, we can neglect the set of intrinsic growth rates $\{r_i\}$ and set $r_i = 1$ for all species (this only changes the rate at which each species reaches its equilibrium yield). In this way equation (3.13) transforms into:

$$\frac{dn_i}{dt} = n_i\left(1 + \frac{\sum_{j=1}^{S}\alpha_{ij}n_j}{K_i}\right) \qquad (\alpha_{ii} = -1)i = 1,\ldots, S. \tag{3.13$'$}$$

So our starting point is the set of equilibrium abundances predicted by equation (3.13), which verify:

$$n_i^*\left(K_i + \sum_{j=1}^{S}\alpha_{ij}n_j^*\right) = 0 \qquad i = 1,\ldots, S, \tag{3.14}$$

where the asterisks denote quantities at equilibrium. An experimental straightforward procedure to estimate these parameters is to perform, during sufficiently long enough periods (in order that the equilibrium state is reached):

(a) the S single species or monoculture experiments, and from each of them to estimate the carrying capacities as the yield of the species i in monoculture m_i^{ex*}(we use m to emphasize that they are yields in monoculture, while the superscript 'ex' is for denoting experimentally measured quantities to distinguish them from the theoretical ones);
(b) the $S\times(S-1)/2$ pairwise experiments and for each of them, obtain the pair of the *biculture* (pairwise experiments) yields, $n_{i(j)}^{ex*}$and$n_{j(i)}^{ex*}$ (the subscripts $i(j)$ and $j(i)$ stand for the relative yield of species i in presence of species j and vice versa).

Therefore, using (a) we obtain K_i, and then from (b) we obtain α_{ij} and α_{ji} by solving equation (3.14) for $S = 2$, as:

$$K_i = m_i^{ex*}, \tag{3.15a}$$

$$\alpha_{ij} = \frac{n_{i(j)}^{ex*} - m_i^{ex*}}{n_{j(i)}^{ex*}}, \qquad \alpha_{ji} = \frac{n_{j(i)}^{ex*} - m_j^{ex*}}{n_{i(j)}^{ex*}}. \tag{3.15b}$$

Therefore, if the yield of species i in biculture with species j is $> K_i$ ($<K_i$) then $\alpha_{ij} > 0$ (<0) and the interaction of j on i is facilitative (competitive). This is the kind of approach followed by Vandermeer (1969) in a pioneering experimental study with protozoa.

Remark: A criticism is that model predictions in general assume that the community is in equilibrium. In the strict mathematical sense a community is at equilibrium only when the rate of change for all species, i.e., the left hand side of equation (3.13) is zero. But this theoretical ideal is rarely achieved in natural communities (Wiens 1984, Roxburgh and Wilson 2000). Moreover, in nature the carrying capacities and the intensity of interactions may vary with a range of environmental factors including climate, kind of resources, spatial distribution of resources and temporal variation in all of the foregoing. Thus, the parameters themselves become variable with time.

3.2.2 Quantifying the accuracy of the linear model for predicting species yields in single-trophic communities[3]

The question we want to answer is: can LLVGE accurately predict the species yields at equilibrium in a community of interacting species?

Based on the previous subsection, to answer this question we will use a dataset of 33 experiments compiled from the literature designed to measure the effects of intra and interspecific interactions in single-trophic communities with $S > 2$ species (Fort 2018). Some of these experiments were completed in the laboratory and others in the field under natural conditions; they included mostly plants but also algae, crustaceans and protozoa. Such experiments measured *all* the yields for the treatments listed below:

(i) the yields of species in monoculture, $\{m_i^{\mathrm{ex}*}\}$, (S treatments),
(ii) the yields of species in bicultures, $\{n_{i(j)}^{\mathrm{ex}}*,\ n_{j(i)}^{\mathrm{ex}}*\}$,($S\times(S-1)/2$ treatments),
(iii) the equilibrium yields of the S coexisting species in polyculture,$\{n_i^{\mathrm{ex}}*\}$, (one treatment).

As explained in the previous subsection, such experiments were carried out until all the species yields versus time seem to stabilize at equilibrium constant values.

Therefore, for each experiment, from (i) and (ii) we can estimate $\{K_i\}$ and $\{\alpha_{ij}\}$ through, respectively, equations (3.15a) and (3.15b). Then feeding the LLVGE with these parameters we will compare the 'theoretical' equilibrium yields they produce, $\{n_i*\}$, against the measured $\{n_i^{ex}*\}$ in (iii). The straightforward way of obtaining the n_i* is by solving the equation (3.14) for the given set of empirically determined parameters $\{K_i,\alpha_{ij}\}$. The equilibrium state in which all the S species coexist reduces equation (3.14) to this simpler equation:

$$\sum_{j=1}^{S}\alpha_{ij}n_j^* = -K_i \qquad i = 1,\ldots, S, \qquad (3.16)$$

[3] This section is based on Fort (2018).

or in matrix form:

$$\mathbf{An}^* = -\mathbf{k}, \tag{3.16'}$$

where **A** denotes the $S \times S$ Lotka–Volterra matrix $[\alpha_{ij}]$, $\mathbf{n}^*$ and **k** are column vectors of S entries with, respectively, the yields n_i and the carrying capacities. And then, by inverting this matrix relationship we obtain the column vector $\mathbf{n}^*$ with the S yields as:

$$\mathbf{n}^* = -\mathbf{A}^{-1}\mathbf{k}. \tag{3.17}$$

Thus, a first required condition to do this inversion is that the matrix **A** is invertible, i.e., $\det\mathbf{A} \neq 0$. All the 33 **A** matrices obtained from biculture yields are invertible. In box 3.1 is shown an example of obtaining the theoretical equilibrium state using equation (3.17).

Box 3.1. A worked example of using equation (3.17) to obtain the equilibrium yields

Here we illustrate how to obtain the equilibrium yields from the experimental data involving four species of winter annuals plants of Rees *et al* (1996)—*Erophila verna* (E), *Cerastiums emidecandrum* (C), *Miyosotis ramosissima* (M), and *Valerianella locusta* (V). We also check the accuracy of predictions and the stability of the equilibrium.

For the area A and year 1979 community the carrying capacity vector **k** and the interaction matrix A are given by:

$$\mathbf{k}^{A79} = \begin{bmatrix} 331 \\ 499 \\ 145 \\ 335 \end{bmatrix} \tag{3.18}$$

$$\mathbf{A}^{A79} = \begin{bmatrix} -1 & 0 & 0 & -0.08 \\ -0.23 & -1 & 0 & 0 \\ -0.01 & 0 & -1 & -0.15 \\ 0 & -0.03 & -0.23 & -1 \end{bmatrix} \tag{3.19}$$

$\det(\mathbf{A}^{A79}) = 0.9675$, and thus the matrix is invertible. Substituting (3.18) and (3.19) into (3.17) we get the vector of equilibrium yields:

$$\mathbf{n}^* = \begin{bmatrix} 307.01 \\ 428.39 \\ 96.95 \\ 299.85 \end{bmatrix}, \tag{3.20}$$

which is consistent with the experimental one (i.e., within the experimental error bars)

$$\mathbf{n}^{ex*} = \begin{bmatrix} 248.2 \pm 139.7 \\ 680.8 \pm 279.2 \\ 135.9 \pm 84.2 \\ 259.5 \pm 91.5 \end{bmatrix}. \tag{3.21}$$

That is, the percentage of predictions which fall within the error bars with 95% confidence, *P*95, is equal to 100%. What about the stability of this equilibrium? It turns out that an equilibrium $\mathbf{n}^*$ is locally stable if all the real parts of the eigenvalues of the Jacobian matrix are negative (appendix). By equations (3.11) and (3.12), the elements of this matrix are given by $J_{ij} = r_i a_{ij} n_i^*$, and, as we are taking $r_i = 1$, $J_{ij} = a_{ij} n_i^*$. Then,

$$J = \begin{bmatrix} -307.01 & 0 & 0 & -25.56 \\ -98.53 & -428.39 & 0 & 0 \\ -0.97 & 0 & -96.95 & -14.54 \\ 0 & -9.00 & -68.97 & -299.85 \end{bmatrix} \tag{3.22}$$

Since the four eigenvalues λ_i of the Jacobian matrix have negative real parts: $\lambda_1 = -0.92$, $\lambda_2 = -4.30$, $\lambda_3 = -3.05 + 0.13i$ and $\lambda_4 = -3.05 - 0.13i$ we conclude that this equilibrium is locally stable.

However, when working with real data, the straightforward procedure to obtain the equilibrium yields exemplified in box 3.1 has problems. In fact, the pairs A and k estimated for several experiments produce via the matrix equation (3.17) negative yields for some species. This means that the pair {**A**,**k**} obtained—by equations (3.15)—is not fully consistent with the equilibrium in which all the species coexist, i.e., this equilibrium is unfeasible and thus $n_l^* = 0$ for some species *l*. Therefore, rather than equation (3.16), which is valid for an equilibrium in which all species coexist, we have to consider the full equilibrium equation (3.14). The problem is that instead of a single equilibrium equation (3.13′) now has multiple equilibria (in which at least one species is extinguished). This theoretical equilibrium with $n_l^* = 0$ would still be consistent with the experimental one provided that the corresponding empirically measured n_l^{ex*} is small enough (so that $n_l^* = 0$ falls within the 95% confidence interval around n_l^{ex*}). To find among this set of possible equilibria the one towards which the system converges we can simulate the dynamical equation (3.13′), starting from a random set of initial values n_i°. It turns out that equation (3.13′) generates for each of the 33 experiments 'theoretical' trajectories $n_i(t)$ for each species that converge to an equilibrium value if the total number of time steps *T* is large enough (at most few hundred time steps). Thus, for each experiment we will compare $n_i(T)$ against the empirical n_i^{ex*} for each species *i*. Figure 3.2 shows that simulating equation (3.13′) for Rees *et al* (1996) the system converges towards the equilibrium state of equation (3.20) for different initial conditions.

Let us present some useful accuracy metrics to quantitatively evaluate the performance of the LLVGE for this set of 33 experiments.

A commonly used metric to assess how well a model fits observed data is the familiar Pearson's correlation coefficient (*r*) or its square, the coefficient of determination (R^2). Nevertheless, a problem with these two statistics is that they actually describe the degree of *collinearity* between the observed and model-predicted values rather than their numerical agreement (Willmott 1984). In fact, by their very definition, both indices are insensitive to additive and proportional differences between the model predictions and observations (Willmott 1984). Thus,

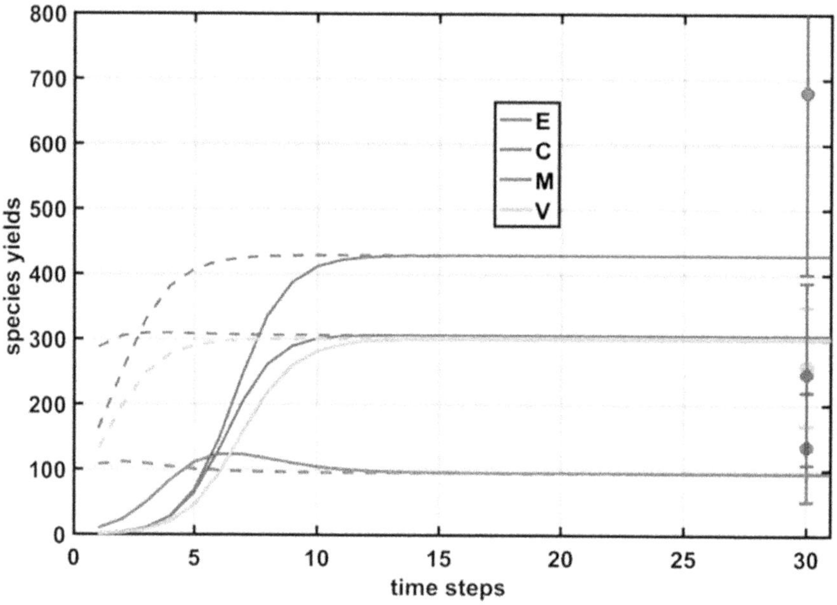

Figure 3.2. Theoretical yields produced by two different simulations, i.e., different initial conditions (filled and dashed lines), converging to experimental values (discs with error bars) for the Rees *et al* (1996) experiments..

both suffer from limitations that make them poor measures of model performance. For further discussion on why r and R^2 are incorrect measures of predictive accuracy we refer the reader to Li (2017) and references therein. Therefore, to quantitatively assess the degree to which the LLVGE match the observations, we will resort to different indices. We summarize the four metrics of goodness-of-fit we will use in table 3.2 (some of them commonly used in atmospheric and hydrologic sciences). **Thereafter, since we always assume an equilibrium or quasi-equilibrium state, in the rest of this chapter we will omit the * for all yields on the understanding that they are yields in equilibrium.**

Notice that all the above metrics are in terms of the absolute value instead of the square of differences. This is because absolute values are preferable over squares since by using absolute values errors differences are given a more appropriate weighting, not inflated by their squared values (Willmott 1981). Squaring in statistics is useful because squares are easier to manipulate mathematically than are absolute values, but use of squares forces an arbitrarily greater influence on the statistic by way of the larger values (Legates and McCabe 1999).

Let us briefly comment on these indices:

The relative mean absolute error (*RMAE*) is obtained from dividing the mean absolute error (*MAE*) between the mean of the species yields. *MAE* and the similar root mean square error (*RMSE*) are two commonly used measures for assessing the predictive accuracy in the environmental sciences (Li and Heap 2008). To avoid any dependence of *MAE* on *S* we will use the relative metric *RMAE*. In order to quantify the accuracy we need to introduce some reference point. Actually error measures,

Table 3.2. Mathematical definitions of error/accuracy measures used in this study. $\overline{n^{\mathrm{ex}}}$ and SE denote, respectively, the mean and standard error for the experimental yields (from Fort 2018).

Error/accuracy measure	Definition
Relative mean absolute error ($RMAE$)	$RMAE = \frac{\sum_{i=1}^{S} \mid n_i^{\mathrm{ex}} - n_i \mid / S}{\overline{n^{\mathrm{ex}}}} 100\ (\%)$
Predictions within 95% confidence intervals ($P95$)	$P95 = \frac{\sum_{i=1}^{S} \delta(\mid n_i^{\mathrm{ex}} - n_i \mid)}{S} 100\ (\%),\quad \delta(x) = \begin{cases} 1 \mathrm{if} x < 1.96\mathrm{SE} \\ 0 \mathrm{if} x > 1.96\mathrm{SE} \end{cases}$
Modified coefficient of efficiency	$E_1 = 1 - \frac{\sum_{i=1}^{S} \mid n_i^{\mathrm{ex}} - n_i \mid}{\sum_{j=1}^{S} \left| n_j^{\mathrm{ex}} - \overline{n^{\mathrm{ex}}} \right|}$
Modified index of agreement	$d_1 = 1 - \frac{\sum_{i=1}^{S} \mid n_i^{\mathrm{ex}} - n_i \mid}{\sum_{j=1}^{S} \left(\left| n_j - \overline{n^{\mathrm{ex}}} \right| + \left| n_j^{\mathrm{ex}} - \overline{n^{\mathrm{ex}}} \right| \right)}$

like $RMAE$, are not accuracy measures, so they can only tell which model produces less error but they are unable to tell how accurate a model is (Li 2017). At any rate, a very tolerant measure of model goodness would be $RMAE < 100\%$, i.e., every quantity is measured with an error smaller than the size of the quantity itself. We will consider here the more stringent threshold of $RMAE < 50\%$. A reference point of 50% might seem too high. Still, it is comparable with the typical SE of the experimental yields.

$P95$ measures the percentage of predictions which fall within the confidence intervals of 1.96σ (within the error bars shown in figure 3.3). $P95 = 100\ (0)\ \%$ means that all (none of) the yields predicted by the model fall within the error bars around the corresponding experimental values. For $P95$ we will consider two thresholds: its maximum possible value of 100% and the (arbitrary) 66.7%, so that $P95 \geqslant 66.7\%$ indicates the model does a decent job.

E_1 (Legates and McCabe 1999) is a modified version of the coefficient of efficiency (Nash and Sutcliffe 1970) defined by $E = 1 - MSE/\sigma^2$ (MSE = mean square error), but in terms of absolute differences rather than square differences. It ranges from minus infinity to 1, the larger its value the better the agreement. In particular, $E_1 = 1$ indicates a perfect match between model predictions and measures. For example, if the absolute differences between the model and the observation are as large as the variability in the observed data (measured by $\sum_{i=1}^{S} \mid n_i^{\mathrm{ex}} - \overline{n^{\mathrm{ex}}} \mid$), then $E_1 = 0.0$, and if it exceeds it, then $E_1 < 0.0$ (i.e., the observed

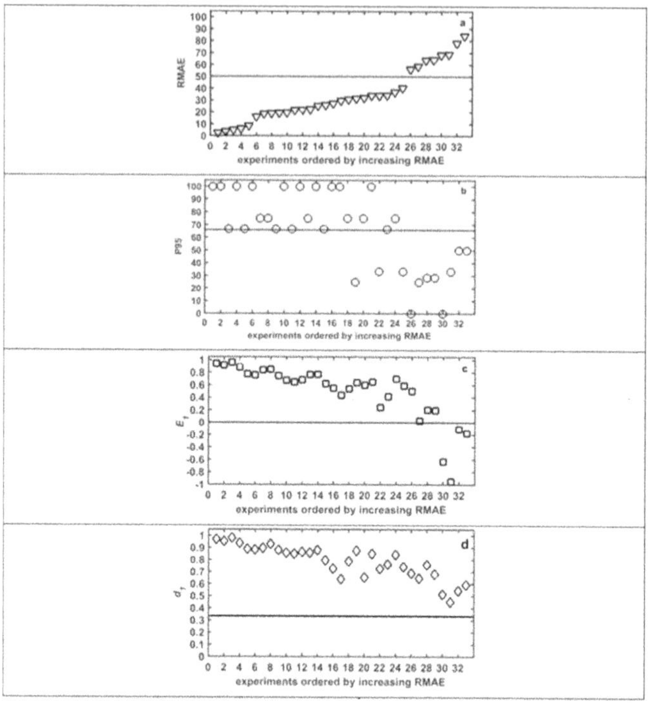

Figure 3.3. The four error/accuracy indices for the 33 experiments of table 3.3 ordered by increasing *RMAE*. Filled lines correspond to the reference values for *RMAE, P95*, E_1, and d_1, distinguishing good from bad performance, respectively: 50%, 66.7%, 0, and 1/3.

mean is a better predictor than n_i). In other words, a value of zero for E_1 indicates that the observed mean $\overline{n^{ex}}$ is as good a predictor as the model, while negative values indicate that the observed mean is a better predictor than the model (Wilcox *et al* 1990).

d_1 (Legates and McCabe 1999) is an index similar to E_1. An advantage over E_1 is that it is bounded, i.e., it varies from 0.0 to 1.0 (again the higher its value the better the agreement between model and observations). By the same token it loses the meaningful reference point of 0.0 for the coefficient of efficiency, which serves to assess when the model is a better predictor than the observed mean. However, it is possible to introduce a reference point for d_1 by observing that for two completely uncorrelated random vectors drawn from a uniform distribution, d_1 is on average = 1/3 (Fort 2020). Therefore if $d_1 \leqslant 1/3$ we conclude that the model is a poor predictor.

The values of the above four error/accuracy metrics for each of the 33 experimental studies are shown in figure 3.3.[4] Filled lines correspond to the reference

[4] The interested reader can find the data for these experimental studies—including interaction matrices **A**, n_i^{ex} and n_i as well as the contribution of each species to the four metrics used for estimating error/accuracy—in supplementary Table S1 of Fort (2018).

values for distinguishing good from bad performance, respectively: 50%, 66.7%, 0, and 1/3. Note that these 33 experiments were ordered from smaller to larger *RMAE*. As expected, when the *RMAE* increases the three accuracy metrics decrease. Thus qualitatively we see that the LLVGE do a good job in predicting yields up to the 25th experiment.

In more quantitative terms, notice that all the 33 experiments have $RMAE <$ 100% (25 of them with $RMAE <$ 50%) and simultaneously $d_1 > 1/3$; 22 have $P95 \geqslant$ 66.67% and 28 have $E_1 > 0$. Furthermore, two thirds of the 33 experiments simultaneously verify the four above inequalities with $RMAE <$ 50% and $P95 \geqslant$ 66.7% (this number reduces to 10 experiments when the maximum $P95 =$ 100% is required).

We have therefore found that by obtaining the generalized Lotka–Volterra parameters as mean values over samples of biculture and monoculture experiments, the LLVGE can accurately predict not all but the majority of the species yields for most of the experiments.

3.3 The statistical mechanics of populations

3.3.1 Rationale and first attempts

As we mentioned, Lotka's program to create the new biological discipline of physical biology included the development of a statistical mechanics approach similar to that of Boltzmann and Gibbs formulated to explain classical thermodynamics from the statistical behavior of atoms or molecules. One of his goals was to understand evolution broadly as a process involving the capture and exchange of energy (Kingsland 2015). Lotka believed that natural selection could be understood as a physical principle with the same level of generality as the laws of thermodynamics (Lotka 1922a). He proposed the principle that 'evolution proceeds in such direction as to make the total energy flux through the system a maximum compatible with the constraints' (Lotka 1922b). In fact, Lotka went further and claimed that the law of evolution coincides with the Second Law of Thermodynamics (Lotka 1925).

Lotka's ideas on the energetics of evolution inspired the American ecologist H T Odum, known for his pioneering work on *ecosystem ecology*—the integrated study of living (biotic) and non-living (abiotic) components of ecosystems and their interactions within an ecosystem framework. Odum proposed the theory that natural systems tend to operate at an efficiency that produces the maximum power output, not the maximum efficiency (Odum and Pinkerton 1955).

This tradition of population dynamics based on maximum principles and in terms of a statistical mechanics formulation continued with Kerner, who proposed the first statistical mechanical formulation of the Lotka–Volterra model (Kerner 1957, 1959, 1972). He used the Boltzmann–Gibbs formulation of statistical mechanics which requires a conserved quantity that remains constant with motion, namely the total energy. Kerner showed that if $\alpha_{ij} = -\alpha_{ij}$ then the quantity

$$\Phi \equiv \sum_{i=1}^{S} [n_i(t) - n_i^* \ln n_i(t)] \tag{3.23}$$

is conserved. This may be verified by differentiating equation (3.23):

$$\frac{d\Phi}{dt} \equiv \sum_{i=1}^{S}\left[\frac{dn_i(t)}{dt} - \frac{n_i^*}{n_i(t)}\frac{dn_i(t)}{dt}\right]. \tag{3.24}$$

Substituting equations (3.13′) and (3.16) into equation (3.24) we get

$$\frac{d\Phi}{dt} \equiv \sum_{i,j=1}^{S}(n_i(t) - n_i^*)\alpha_{ij}\big(n_j(t) - n_j^*\big). \tag{3.25}$$

It is immediate to see that the rhs of equation (3.25) vanishes if the Lotka–Volterra matrix $[\alpha_{ij}]$ is antisymmetric. This is an important limitation of Kerner's formulation since it cannot accommodate either mutually intraspecific competition or competitive or mutualistic interspecific interactions.

Goel *et al* (1971) tried to obtain salient properties of populations which are rather insensitive to initial conditions and details of assumptions, and thus amenable to a statistical mechanics approach. The authors first analyze the case in which the interactions between various species and their growth coefficients are known, and show that *a priori* one can determine whether the population will be stable or not and, if not, which of the species will disappear. Also, they discuss the stability of the population when several foreign species are introduced. Then Goel *et al* move to the situation when the information about the detailed interactions between various species is absent, as is usually the case; thus a statistical mechanical treatment of the population is desirable. They start with the dynamics of the system described by the Lotka–Volterra equations and determine the conditions under which the canonical ensemble averages (see section 2.1.4) satisfy the equations by the time averages of arbitrary functions of a number of various species. Goel *et al* show that for small deviations from steady state populations, **the necessary and sufficient condition for such a treatment is that the number of species is large**. However, for arbitrary deviations the latter is a necessary condition and may not be sufficient, thus compromising the justification of the use of statistical mechanics (at least from the viewpoint of the classical Boltzmann–Gibbs formulation). At any event, this statistical mechanical treatment provides an empirical method for measuring the stability of an ecosystem and the effect of changes in the interactions between various species, due to changes in temperature, humidity, age distribution, and other ecological abiotic factors.

3.3.2 MaxEnt formulations

As we have seen in chapter 2, Jaynes' MaxEnt has a wider applicability than either Boltzmann or Gibbs ensemble formulation of statistical mechanics since it neither requires the postulation of the existence of a Liouville equation, nor of a conserved quantity, nor of an ergodic hypothesis. Thus, subsequently the MaxEnt formulation of population dynamics was considered by several authors. In this subsection we want just to provide a non-exhaustive list of such MaxEnt treatments, which includes: Hamann and Bianchi (1970), Alexeyev and Levich (1997), Shipley *et al*

(2006), Pueyo *et al* (2007), Harte *et al* (2008), Volkov *et al* (2009), Banavar *et al* (2010), He (2010) and Harte (2011).

In chapter 5 we will consider a MaxEnt-based formulation of ecological communities and use it to estimate effective interaction coefficients between different entities or agents, like firms, species, etc. Then, this procedure will allow us to address practical problems in chapters 6 and 7 to infer the Lotka–Volterra interaction matrix for describing the dynamics of two different ecosystems. In chapter 6 we will analyze the dynamics of markets, both qualitatively and quantitatively. In chapter 7 we will apply this method to forecast the trajectories of tree species in tropical forests.

3.3.3 Uses of statistical mechanics-inspired lattice models I: overgrazing of semi-arid lands

Since the pioneering works of Lotka, Odum and Kerner attempting to formulate ecology in terms of a statistical mechanics, a growing number of statistical mechanics formulations of population dynamics have been proposed.

In this subsection we will show how, despite the formal limitations of the straightforward statistical mechanics formulation of ecosystems, statistical mechanics-inspired models allow extending the Lotka–Volterra dynamical equations to spatially explicit descriptions of ecosystems. That is, the Lotka/Volterra equations provide what is called a *mean field* (MF) description. MF models neglect all spatial heterogeneities and describe the change over time of some averaged variable over space. Now we will introduce space variability. A simplified way to introduce spatial degrees of freedom is through **lattice modelling**, i.e., replacing the continuous ordinary space by a lattice, a common useful approach in statistical mechanics. A known example is provided by the spin models of magnetism, like the Ising model we studied in chapter 2. different from the models considered in the previous subsection, for which the growth rate of population of each species is given in terms of the intraspecific and interspecific interaction, lattice models are 'microscopic'; they are formulated in terms of the trait of each individual of a species, such as viability, fecundity, motility, and social behaviors. In this sense their treatment is fully statistical mechanical. Lattice models are particularly useful because they are amenable to computer simulation and are suited to performing *virtual experiments* to get insight into real-world problems.

The first lattice model we will consider corresponds to one species that is being harvested. To fix ideas imagine a population of plants whose biomass dynamics, in absence of herbivores, would be governed by logistic equation (3.5). The effect of herbivory is usually modeled through a sigmoid consumption term (Noy-Meir 1975) in such a way that it leads to this model in terms of a non-dimensional variable X for the plant biomass density (Fort 2020):

$$\frac{dX}{dt} = X\left(1 - \frac{X}{K}\right) - c\frac{X^2}{X^2 + 1}, \tag{3.26}$$

in terms of two also non-dimensional parameters; the carrying capacity, K, and the maximum consumption rate or *grazing pressure*, c.

Equation (3.26) is known to have for some range of both parameters multiple equilibria (Fort 2020), i.e., under the same external conditions, the system can be different equilibrium states which are also called ***alternative stable states*** (ASS) (Carpenter 2001). The appearance of ASS is mathematically equivalent to say that the system experiences a ***bifurcation***. Hence, when subjected to a slowly changing external factor (such as climate or human activities), an ecosystem may show little change until it reaches a critical point where a sudden shift to an alternative contrasting state occurs.

Figure 3.4 depicts the region in the plane c–K inside of which we have ASS, which is delimited by the bifurcation set S_B (black solid lines). In the region to the left of the 'wedge' we have only one equilibrium or stable state of 'high' vegetation density. Inside this region of the parameter space we have two alternative stable vegetation equilibria: in addition to the stable 'high' vegetation equilibrium there are two additional equilibria; a stable 'low' vegetation equilibrium and an unstable equilibrium with an intermediate value of vegetation density. Finally, in the region to the right we have only one equilibrium or stable state of low vegetation density.

Therefore, equation (3.26) serves to describe the catastrophic shift phenomenon of desertification of drylands subject to overgrazing (Adeel *et al* 2005). Overgrazing is regarded as one of the major causes of desertification (Dregne 1986, Wiesmeier 2015). As a result of the grazing pressure the vegetation density decreases. Imagine

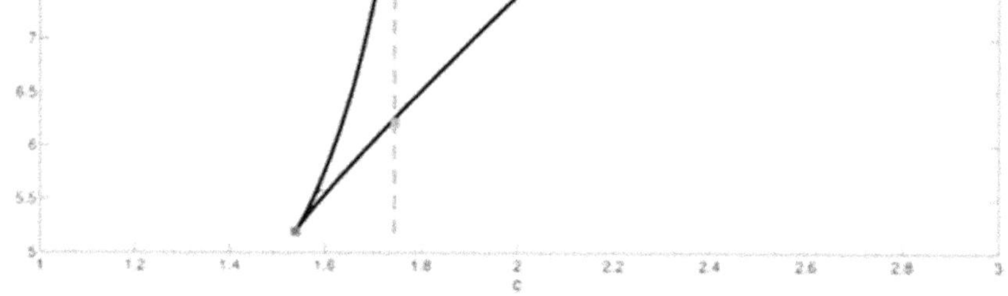

Figure 3.4. The parameter plane of the *grazing* model equation (3.26). The bifurcation set S_B (black solid lines), is a ***separatrix***, i.e., the lines separate the region in which there are three alternative stable states—two of them stable, and one unstable—from the rest of the plane where there is only a single stable state. The cusp point occurs at $c_c = 8/3^{3/2}$, $Kc = 3^{3/2}$ (Fort 2020). If c is varied from outside the wedge region with 3 ASS along the red dashed horizontal line at $K = 7.5$, at the points indicating with red squares a bifurcation occurs. Likewise, if K is varied from outside the wedge region with 3 ASS along the cyan dashed vertical line at $c = 1.75$, at the points indicating with cyan squares a bifurcation occurs.

for instance that for $K = 7.5$, c is increased from a point at the left of S_B. At first, increasing the grazing pressure has no dramatic consequences. When we intersect the left border of S_B a bifurcation occurs. An additional stable 'low' vegetation density equilibrium appears out of the blue. However, the system remains in its upper equilibrium branch without providing any significant warning signal (we only observe a gradual decrease in the vegetation density). If we continue increasing the grazing pressure, when we intersect the right border of S_B the 'high' vegetation equilibrium disappears. Thus, at this point, a sudden steep drop of the vegetation density towards the only available stable 'low' vegetation equilibrium takes place.

We want to analyze the effect of spatial heterogeneities and spatial fluctuation in the desertification process as well as to compute several related metrics of interest involving vegetation spatial distribution. With this aim we will transform the non-spatial differential equation (3.26), into a two-dimensional lattice difference equation (i.e., both space and time discrete). Hence, we represent space by a $L \times L$ regular square lattice so that on each cell of side a, centered at coordinates ($x = a.i$, $y = a.j$) with i and j integer numbers, a plant biomass density is assigned. A straightforward lattice version of equation (3.26) can thus be written as:

$$X(i, j; t+1) = X(i, j; t) + X(i, j; t)\left(1 - \frac{X(i, j; t)}{K(i, j)}\right) - c(t)\frac{X(i, j; t)^2}{1 + X(i, j; t)^2} +,$$

$$d(X(i+1, j; t) + X(i-1, j; t) + X(i, j+1; t) + X(i, j+1; t) - 4X(x, y; t)), \tag{3.27}$$

where the carrying capacity $K(i,j)$ is a local, i.e., spatially heterogeneous, parameter that varies from point to point, the parameter $c(t)$ is taken as uniform but time-dependent, and d is a diffusion coefficient (to allow the spreading of the plant biomass).

Equation (3.27) defines a **cellular automaton** (CA) (Wolfram 1994). A CA is a collection of 'colored' cells on a grid of specified shape that evolves through a number of discrete time steps according to a set of rules based on the states of neighboring cells. The rules are then applied iteratively for as many time steps as desired. Cellular automata were introduced in the 1940s by Stanislaw Ulam and John von Neumann as a model of self-replicating systems. Cellular automata were studied in the early 1950s as a possible model for biological systems (Wolfram 2002).

In our case, the cells of the CA can represent an elementary quadrat, defining the resolution with which we want to describe a land region (figure 3.3). For example, in a recent study of the African Sahel region, the resolution imposed by the remote sensing limitations of satellite images was 30 m × 30 m (Weissmann and Shnerb 2016). Thus, depending on the size we want to cover, we can use a = 30 m, 60 m, 300 m, etc. Likewise, when working with real data, commonly used vegetation indices as proxies of the vegetation density $X(i,j)$ are NDVI (normalized difference vegetation index) or the more reliable EVI (enhanced vegetation index), that incorporates corrections to both soil reflectance and atmospheric disturbances (Weissmann and Shnerb 2016). In the greener the cell the higher the vegetation density in the corresponding quadrat.

In addition to the grid on which a cellular automaton lives and the colors its cells may assume, the neighborhood over which cells affect one another must also be specified. The simplest choice is 'nearest neighbors,' in which only cells directly adjacent to a given cell may be affected at each time step. This is the so-called von Neumann neighborhood (the four cells highlighted in red surrounding the cell marked with an '**x**' in figure 3.5), which was our choice (notice that the diffusion term of equation (3.27) includes these four cells). Box 3.2 summarizes the properties defining the CA. Therefore, we start with an initial random assignation of biomass densities, $X(i,j;1)$ and equation (3.27) provides the CA 'updating rule' which generates the subsequent $X(i,j;t)$ for $t > 1$ (box 3.2).

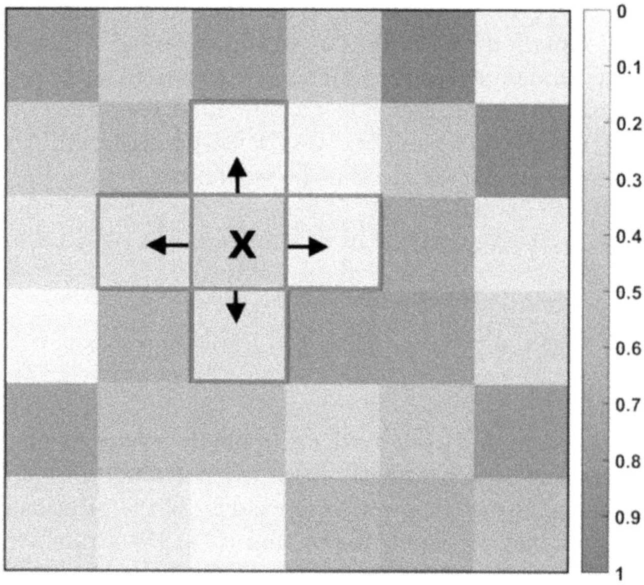

Figure 3.5. A portion of a cellular automaton. In this plot each cell has a normalized vegetation density varying between 0 and 1. The cell marked with 'x' interacts with its von Neumann neighborhood (highlighted in red).

Box 3.2. Properties, variables and parameters of the CA implementing the explicitly spatial version of the grazing model

- Shape of lattice: square.
- Size L: ranging from 100 to 1000 (in fact, for different values of L in this range, no important differences were found).
- Boundary conditions: periodic boundary conditions (PBC). That is, the cells at the border ($i = 1$ or L or/and $j = 1$ or L) only have two neighbors instead of four. One way to avoid artifacts due to border effects is to assume that cells at $i = 1$ and $i = L$ are neighbors (and the same for $j = 1$ and $j = L$). This is equivalent to curling the square into a cylinder, and then joining the cylinder's circular ends to each other, yielding a torus or donut-shape form.

- Cell variable: non-dimensional vegetation density $X(i,j;t)$.
- Parameters and their ranges: The ranges of values for the parameters that we use are chosen to contain the region of alternative stable states determined in equation (3.26): the carrying capacity $K(i,j)$ varies randomly from cell to cell around a fixed spatial mean $\langle K\rangle = 7.5$ in the interval $[-\delta K, \delta K]$ where $\delta K = 1.0$–2.5. Typical values for the consumption rate c are between 1 and 3. The diffusion parameter d is taken varying between 0.1 and 0.5.
- Neighborhood: von Neumann neighborhood, i.e., each cell is connected to its four nearest neighbors.
- Updating rule: equation (3.27).
- Number of time steps T: The number of iterations is typically 1000.

The spatial metrics informative of the desertification shift we compute are:
The spatial variance σ^2_X:

$$\sigma_X^2 = \langle X(t)^2\rangle - \langle X(t)\rangle^2, \tag{3.28}$$

where $\langle\rangle$ denote the *spatial mean* at fixed time t, i.e., the sum of the variable over all cells.

- The *patchiness* or *cluster structure*. Clusters of high (low) X are defined as connected regions of cells with $X(i, j, t) > X_m$ ($X(i, j, t) < X_m$) where X_m is a threshold value. There are different criteria for defining X_m (see below).
- The *two-point correlation function* for pairs of cells at (i_1, j_1) and (i_2, j_2), separated by a given distance R, which is given by:

$$G2(R) = \left\langle X(i_1, j_1)X(i_2, j_2)\right\rangle - \left\langle X(i_1, j_1)\right\rangle\left\langle X(i_2, j_2)\right\rangle \tag{3.29}$$

Figure 3.6 shows how the above metrics change when gradually increasing stress on the system, varying c from 1 to 3 in 1000 steps of 0.002, i.e., each 'measure' is performed for a different value of the control parameter c. Panel (a) includes a plot of $\langle X\rangle$ and σ^2_X for a fixed average capacity $\langle K\rangle = 7.5$, $d = 0.1$ and the initial condition for each $X(i,j)$ uniformly randomly chosen in the interval $[0, \langle K\rangle]$. Notice that the spatial variance has a clear peak at $c_m \cong 2.08$. The temporal variance σ^2_t is also included to show that the peak of σ^2_X occurs before the one of σ^2_t (around 110 time steps earlier).

Panel (b) depicts the two-point correlation for distances $R = 0, 1, 2, 3$. Indeed, the spatial variance is a particular case (distance $R = 0$) of the two-point correlation function defined by equation (3.29). Notice that the peak of the correlation for any R occurs at nearly the same value of the control parameter $c \approx c_m = 2.08$.

To study the cluster structure first we have to define a threshold X_m which separates 'low' from 'high' density vegetation cells. For example, for $\langle K\rangle = 7.5$ and $d = 0.1$ a possible choice is to take X_m coinciding with the average value of $X(i,j)$ at $c = c_m$, i.e., $X_m = \langle X\rangle(c_m) \cong 2.89$. This is the threshold we will take. In the first column of panel (c) we include snapshots of typical patch configurations for $c = c_m - 0.1$, $c = c_m$ and $c = c_m + 0.1$. Densely vegetated patches correspond to clusters of cells which are all between yellow and red. Conversely, bare patches

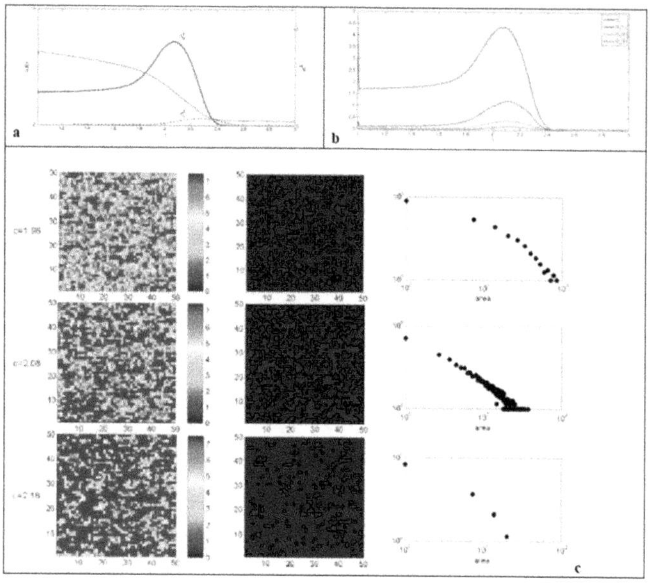

Figure 3.6. Three spatial properties (all of them computed for $\langle K \rangle = 7.5$). (a) $\langle X \rangle$, σ^2_X and, for comparison, the temporal variance σ^2_t (dashed line) for $d = 0.1$, $\langle K \rangle = 7.5$. The peak of σ^2_X occurs at $c_m \cong 2.08$ and the peak of σ^2_t at $c \cong 2.30$. (b) The two-point correlation function for different lengths R, $d = 0.1$, $\langle K \rangle = 7.5$. The peak of $G_2(R)$ for different values of R also occurs at $c_m \cong 2.08$. (c) Patchiness. *First column*: a portion of 50×50 cells from the original 800×800 lattice is shown, grids representing the value taken by $X(i, j)$ at each cell for $\langle K \rangle = 7.5$, $d = 0.1$. The rows correspond to $c = 1.98$, $c = 2.08$ and $c = 2.18$. *Second column*: same as the first, for binarized data (blue: cells with low vegetation density, red: cells with high vegetation density). *Third column*: number of clusters *versus* area on a logarithmic scale.

appear in blue. The second column a binary representation, i.e., dark red (blue) cells correspond to cells for which $X > \langle X \rangle\ c_m$ ($X < \langle X \rangle\ c_m$). The plots in the third column are the corresponding cluster distributions. At $c = c_m$ the patch-size distribution follows a power law over two decades—with exponent $\gamma \approx -1.1$ for $d = 0.1$ and $\gamma \approx -0.9$ for $d = 0.5$—which disappears for smaller or greater values of c. The singular behavior of these three quantities at the same value of c is commonly taken as a signature of an ongoing shift towards desertification. And serve as early warnings of catastrophic shifts of the system (Fort 2013).

Interestingly, the desertification transition has remarkable similarities with one of the most widely analyzed phase transitions in physics: the boiling of a liquid. This is because the grazing model (3.26) can be mapped onto the simplest state equation able to account for the liquid–vapor phase transition, the ***van der Waals equation of state***. This correspondence allows further insights on the phenomenology of catastrophic shifts as well as to connect the behavior of typical observables, like the spatial variance, the two-point correlation function and the patchiness. The idea is that the high and low vegetation density states are, respectively, analogous to the high and low densities of water (i.e., the liquid and gas phases). Thus, the desertification process through the appearance of empty patches can be understood as the nucleation and growth of droplets in liquid water. We refer the interested reader to chapter 4 of Fort (2020).

3.3.4 Uses of statistical mechanics-inspired lattice models II: modelling the dynamics of biodiversity of communities of trees in tropical forests

There are several diversity metrics and relationships to study and monitor biodiversity. The most obvious measure of biodiversity is the *species richness* or the number of species, S, coexisting in the community. Another commonly used metric is the Shannon *equitability* index

$$\boldsymbol{E} = -\sum_{i=1}^{S} \frac{N_i}{N} \log\left(\frac{N_i}{N}\right) / \log(S), \tag{3.30}$$

where N_i is the number of individuals of species i and N is the total number of individuals. Notice that E is simply the normalized Shannon entropy between 0 and 1. In addition to S and E we will consider two popular relationships. First, the ***species-abundance distribution*** (**SAD**), i.e., the probability that in a community with S species and N individuals a species has abundance N_i, also provides useful information. A frequent way to represent the SAD is by providing empirical frequency distributions in the form of rank–species abundance (RSA) relationships. Second, the ***species–area relationship*** (**SAR**), i.e., the average number of species in a region of area A if there are S_0 species in a larger area A_0.

In this subsection we will introduce another Lotka–Volterra lattice model or cellular automaton to obtain the SAD and SAR for communities of trees of in different tropical forests. In fact, our departure point will be the classical competition niche theory, based on the combination of the Lotka–Volterra competition equations and ***Hutchinson's multidimensional niche*** (Hutchinson 1957). The dimensions of the Hutchinsonian niche can be thought as environmental conditions and resources (e.g., light, nutrients, etc), that define the requirements of individuals of a species for its population to persist.

Adopting a parsimonious approach—the simplest model with the least assumptions and variables but with greatest explanatory power—we consider the species distributed along a continuum resource spectrum which can be represented by a hypothetical one-dimensional niche axis (May 1974). To fix ideas we can think on bird species that feed on seeds. As shown in figure 3.7, the beak size of a bird species determines the utilization ability of resources (seeds): for a given beak size, a bird species can optimally feed on a particular size of seed, and its feeding ability drops off for seeds that depart from this size. Therefore, one may consider the niche axis as a gradient that is related to the size of organisms. More generally, species are characterized by 'utilization functions', which describe the ability to exploit a particular resource as it varies along a niche axis (Nee and Colegrave 2006). The idea is that the competence for resources (seeds) between two species is given by the amount of *niche overlap* between them. For example the niche overlap between species A and C is negligible and so it is the strength of the competition between them.

The Lotka–Volterra equations (in the continuum) for S species are given by equation (3.13) with all the interaction coefficients $\alpha_{ij} < 0$ since we are considering that the interspecific interaction between species is competition (i.e., we are neglecting all facilitative interaction). To mathematically implement this notion of competition

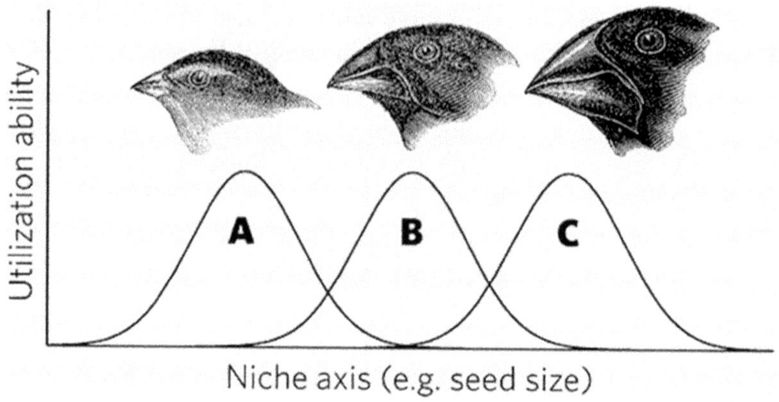

Figure 3.7. A schematic representation of the 'niche space'. Species A, B and C are seed-eating birds in an environment that contains seeds of various sizes. For a given beak size, a bird species can optimally feed on a particular size of seed, and its feeding ability drops off for seeds that depart from this size. More generally, species are characterized by 'utilization functions', which describe the ability to exploit a particular resource as it varies along a niche axis. Reprinted from Nee and Colegrave (2006), copyright 2006 Nature Publishing Group, with permission of Springer.

strength proportional to niche overlap it is customary to assign to each species a position μ_i along a normalized niche axis ξ, between 0 and 1. The utilization ability of each species is then modeled by a normal distribution centered at μ_i with standard deviation σ :

$$P(\xi) = (1/\sqrt{2\pi})e^{-(\xi-\mu)^2/(2\sigma^2)}. \tag{3.31}$$

Therefore, the competition intensity between species i and species j, α_{ij}, is taken, according to the MacArthur and Levins (1967) niche overlap formula, as proportional to $\int_{-\infty}^{+\infty} P_i(\xi)P_j(\xi)\mathrm{d}\xi$.

We consider several simplifications to build a cellular automaton model. We enumerate them below.

Niche properties:

(N_1) The niche space is assumed one-dimensional, and, for simplicity circular ($\xi = 0$ coincides with $\xi = 1$).

(N_2) The set of positions of the species niches, $\{\mu_i\}$, is chosen randomly by drawing the S values from a uniform distribution at the beginning of a simulation (and do not change during the simulation).

(N_3) All species have the same niche width: $\sigma_i = \sigma$ (which could be regarded as an average niche width).

Spatial properties:

(S_1) The system is *saturated*, i.e., all $L \times L$ sites are occupied by one tree (i.e., the number of individuals, $N = L \times L$, remains constant and there are no empty sites).

(S_2) The entire community is considered closed to dispersal from the outside and (as in the previous subsection) we consider periodic boundary conditions to avoid border effects.

(S_3) The dispersal of trees from outside each neighborhood is included through a dispersal rate m.

(S_4) Each focal individual, with integer coordinates (k,l) only interacts with its eight immediate ones, this is the so-called *Moore neighborhood* $M(k,l)$ (figure 3.8). This is of course a simplification, since the effective neighborhood size of tropical trees on average involves a larger numbers of neighbors (Hubbell *et al* 2001) but it varies significantly with the focal species (Uriarte *et al* 2004).

Therefore, at each lattice site (k,l), for a time step t (time is discrete), we have a tree represented by a niche variable $\mu(k,l;t)$ coinciding with one of the $\{\mu_i\}$, i.e., a tree belonging to a given species $i = 1, 2,\ldots,S$ (see below how the initial configuration is chosen and how it subsequently evolves).

By virtue of the MacArthur and Levins (1967) niche overlap formula and conditions (N_1)–(N_3) and (S_1)–(N_3), the strength of the competition for each pair of trees of species i and j located on lattice sites (k,l) and (k',l') is given by:

$$\alpha_{ij}(k, l \mid k', l') = -\frac{\int_0^1 P_i(\xi)P_j(\xi)\mathrm{d}\xi}{0.5*\left(\int_0^1 P_i^2(\xi)\mathrm{d}\xi + \int_0^1 P_j^2(\xi)\mathrm{d}\xi\right)} N(k, l \mid k', l') \tag{3.32}$$
$$= -\mathrm{e}^{-\frac{(\mu_i-\mu_j)^2}{4\sigma^2}} N(k, l \mid k', l'),$$

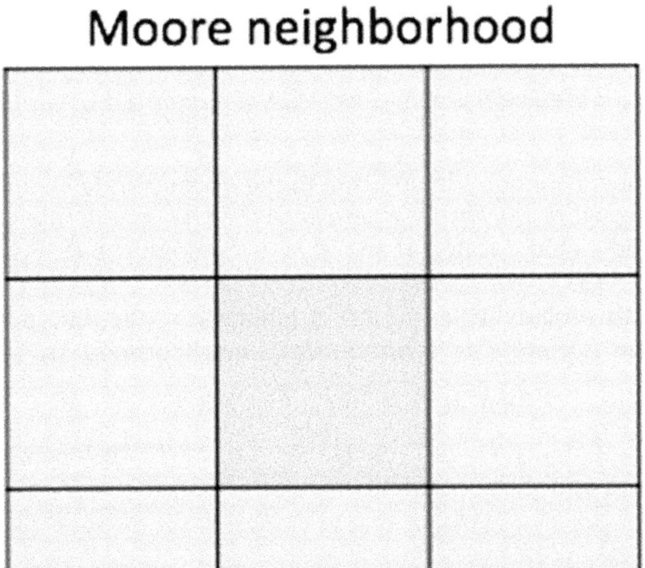

Figure 3.8. The Moore neighborhood used in the CA of this subsection.

with

$$N(k, l \mid k', l') = \begin{cases} 1 \text{if} \mid k - k'\mid_{\text{PBC}} \leqslant 1 \text{and} \mid l - l'\mid_{\text{PBC}} \leqslant 1 \\ 0 \text{otherwise} \end{cases},$$

where'PBC'denotesthatsites$k = 1$ and$k = N - 1$are nearest neighbors.

To implement the Lotka–Volterra competition, a focal node (k,l) is randomly chosen with probability $1/L^2$ per time step. The tree at this node has, as a result of its competition with the other eight individuals in its neighborhood, an effective fitness

$$f(k, l) = 8 + \sum_{k' \neq k, l' \neq l} \alpha_{ij}(k, l \mid k', l'), \tag{3.33}$$

where the '8' corresponds to the numbers of neighbors of (k,l) and it ensures that $f(k, l)$ is always non-negative[5]. (Notice that the sum in equation (3.33) runs over all the sites of the Moore neighborhood of the focal node (k,l), excluding itself, i.e., there are no self-interactions of a tree over itself.)

The initial conditions and the dynamics are as follows.

Initial conditions. Since the initial spatial distribution of individuals at a given forest is not known, we run the model from 100 different initial spatial configurations corresponding to the maximum Shannon equitability. That is, each initial configuration is generated by assigning to each site an individual of a species whose identity was randomly drawn from a uniform distribution so that each species i has a probability $1/S$ of being chosen. Thus, for each simulation there are two draws: this one, to assign to each lattice site a given species, and that of (A) for the species niche positions $\{\mu_i\}$.

- *Dynamics.* This model is a *stochastic* CA (Durrett and Levin 1998, Silvertown *et al* 1992).

Indeed the model contains, in addition to the dispersal rate m, another source of stochasticity through a *stochasticity parameter T*. The update rule is as follows:

(I) A focal individual, located at site (k,l), is randomly chosen (with probability $1/L \times L$).

(II) Then with probability m $\mu(k,l;t)$ is replaced by the descendant of another randomly chosen tree $\mu(k',l';t)$ from outside its neighborhood M_i. i.e.,

$$\Pr(\mu(k, l; t) \to s\mu(k', l'; t)) = m. \tag{3.34a}$$

And, with probability $(1-m)$ a neighbor site (k',l') (belonging to $M(k,l)$) is randomly chosen, and with probability

$$\begin{aligned} &\Pr(\mu(k, l; t) \to \mu(k', l'; t)) \\ &\quad = (1 - m)\{1 + \exp[-(f(k, l; t) - f(k', l'; t))/T]\}^{-1}, \end{aligned} \tag{3.34b}$$

[5] Thus, $f(k,l)$ has its maximum value when the focal species has minimal overlap with its eight neighbours, and minimal when their niche overlaps are maximal.

the individual at (k,l) is replaced by the one at (k',l'). This is the so-called Glauber (1963) *update rule*, commonly used in lattice spin models. Thus, the probability that the focal tree $\mu(k,l;t)$ is replaced by a neighbor tree $\mu(k',l';t)$ increases as the difference between their fitness increases. So, the parameter T is a truly stochasticity parameter: As T approaches its minimum value of 0, $\Pr(\mu(k,l;t) \rightarrow \mu(k',l';t))$ approaches 1 and the replacement of the focal individual by its neighbor becomes the deterministic 'copy the fittest' rule (i.e., the replacement is accepted if and only if $f(k,l;t) < f(k',l';t)$. In contrast, as T approaches ∞, $\Pr(\mu(k,l;t) \rightarrow \mu(k',l';t))$ approaches ½ i.e., the replacement of the focal individual by a neighbor is a totally random process, like flipping a fair coin.

Box 3.3 summarizes the properties and parameters of the CA we consider in this section.

Let us use this CA to explain the empirically observed spatio-temporal dynamics of tree communities using the databases of the Forest Global Earth Observatory (ForestGEO), a global network of scientists and forest research sites dedicated to advancing long-term study of the world's forests. These databases comprise 72 forest research sites across the Americas, Africa, Asia, Europe, and Oceania. Each ForestGEO site is typically between 25 and 50 hectares. In each site, all free-standing trees with a trunk *diameter at breast height* (dbh) of at least 1 cm are tagged, measured, and identified to species. Specifically, we select the nine permanent plots of tropical forests in eight different countries which are listed in table 3.3. Therefore, for each of these plots, L is chosen in such a way that $L \times L$ is the closest multiple of ten to the maximum number of trees (with dbh $\geqslant$ 1 cm) measured along the different censuses, $N_e^{\max}$, while the initial number of species, S, is set equal to the value of the species richness found for the first census, $S_e^{(1)}$ (the subscript e stands for 'empirical' and then $X_e^{(k)}$ will denote the observed value for quantity X in census number k). For example, for Barro Colorado plot ($N_e^{\max} = N_e^{(3)} = 244{,}080$ for the third census of 1990 and $S_e^{(1)} = 320$ for 1982): $L = 500$ and $S = 320$.

Box 3.3. Properties, variables and parameters of the CA implementing the Fort–Inchausti model (2012) of tropical forests.

- Shape of lattice: square.
- Size L: depends on the maximum number of trees of the forest; it ranges from 282 to 600 (see table 3.3).
- Boundary conditions: periodic boundary conditions (PBC).
- Cell variable: position along the one-dimensional niche axis: $\mu(k,l;t)$.
- Parameters and their ranges: the parameters controlling the dynamics are three: the niche width σ, the dispersal rate m, and the *stochasticity parameter* T. The three values vary from forest to forest (see table 3.3).
- Neighborhood: Moore neighborhood, i.e., each cell is connected to its eight nearest neighbors.
- Updating rule: equations (3.34a) and (3.34b).
- Number of time steps T_s: varies with the forest; proportional to L^2.

Table 3.3. The nine tropical forests analyzed. $E_e^{(1)}$ is the empirically measured Shannon equitability index for the first census in each forest (see text); L is the lattice size, σ and T the CA parameters controlling its dynamics which provide the best fit to empirical data, respectively, the species' niche width and the stochasticity parameter; S denotes the empirically observed and the predicted species richness for all trees with diameter at breast height (dbh) > 1 cm.

Forest	Area (ha)	$\mathcal{E}_e^{(1)}$	L	σ, m, T	S	
Lambir (Malaysia)	52	0.863	600	0.073, 0.05, 0.0	Empiric	1204, 1159
					Predict	1204, **1160±9**
Pasoh (Malaysia)	50	0.842	580	0.085, 0.11, 0.0	Empiric	823, 819, 811, 808
					Predict	823, **821±2, 815±4, 808±5**
Yasuni (Ecuador)	50	0.83	389	0.076, 0.13, 2.5	Empiric	1154, 1087
					Predict	1154, **1091±8**
La Planada (Colombia)	25	0.745	337	0.083; 0.09, 1.0	Empiric	241, 221
					Predict	241, **222±7**
Sinharaja (Sri Lanka)	25	0.742	448	0.076, 0.08, 0.0	Empiric	207, 205
					Predict	207, **205±2**
Korup (Cameroon)	50	0.718	574	0.072, 0.13, 0.3	Empiric	495
					Predict	495
Barro Colorado (Panamá)	50	0.694	500	0.077, 0.10, 3.0	Empiric	320, 307, 303, 299, 292, 283
					Predict	320, **314±4, 300±5, 293±6, 287±7, 281±7**
Fushan (Taiwan)	25	0.692	338	0.085, 0.09, 0.0	Empiric	110
					Predict	110
Huai Kha Kaengh (Thailand)	50	0.682	282	0.086, 0.12, 0.0	Empiric	295
					Predict	295

Data from the Forest Global Earth Observatory (ForestGEO 2021).

Computation of biodiversity metrics

The RSA was calculated by counting the number of species falling in each ranked abundance interval. The SAR (average number of species versus plot area) curves were calculated by dividing the entire plot into non-overlapping quadrats of square and rectangular shapes and the number of species present in each counted. We considered quadrats of the following sizes: 5×5, 5×10, 10×10, 10×20, 20×25, 25×25, 50×25, 50×50, 100×50, 100×100, 250×100, 250×250 and 500×500, containing from 25 to 250,000 individuals because all sites are occupied in the model. The mean number of species in quadrats of different sizes yielded the SAR.

We use a method based on a sequential procedure to estimate the parameters σ, m and T providing the best fit to the observed dynamics as characterized by a set of common metrics in the set of censuses of each forest (figure 3.9). Actually, it turns out that while σ and m were enough to reproduce with accuracy the values of all biodiversity metrics found at the first census of each forest plot, T was a 'fine tuning parameter' required to improve the agreement between observed and predicted values of the RSA and the Shannon equitability index E calculated for subsequent censuses. For this reason $T = 0$ for all the forest plots for which there is only one census

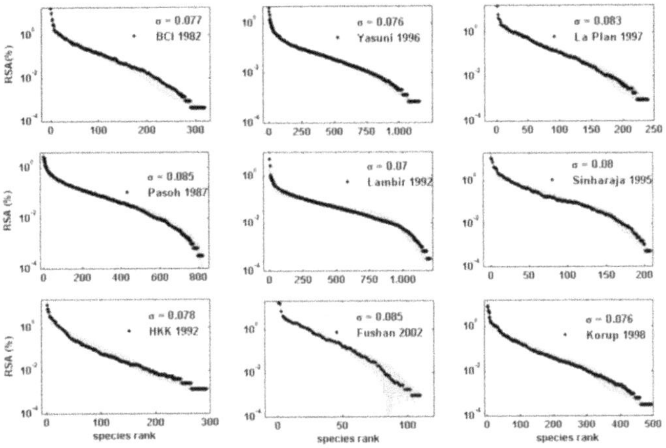

Figure 3.9. Observed (bold) and predicted distribution of relative species abundances (RSA) for all trees with diameter at breast height dbh $\geqslant$ 1 cm for the first census of nine in nine tropical forest plots. Data from ForestGEO (2021). Averages over 100 model simulations ± std for the best estimates of model parameters of each forest (see text).

(see table 3.3). Thus in a first stage, using only data of the first census, we estimated σ and m. To do this we systematically searched the array of values in the plane σ–m generated by varying σ in [0.05, 0.1] in steps of $\Delta\sigma = 0.001$ and m in [0.02, 0.12] in steps of $\Delta m = 0.01$. However, we do not know *a priori* how many simulation steps are required to yield a configuration comparable to the one observed in the first census starting from random initial conditions. The recipe we use is, for each pair (σ, m), we generate 100 initial conditions (see above) and we run each simulation until the predicted value of E is equal to the empirical $E_e^{(1)}$ for the first census with an accuracy of 1%. At this point we stop the simulation[6]. In this way we arrive at several possible pairs (σ', m'); to select the best estimate we choose the pair (σ, m) that yields the highest coefficient of determination R_{et}^2 between the observed and predicted (average over the 100 simulations) RSA distributions, and simultaneously verifies $R_{et}^2 \geqslant 0.95$.

In a second stage, the pair of fitted (σ, m) is then used to estimate the other parameter, T. To do this we have to restart the simulation at each CA configuration corresponding to the first census but now allowing species to become extinct so that the predicted forest dynamics is able to describe the observed changes in S for subsequent censuses of each forest. Proceeding in a similar way as before, we systematically search for the best fitting value of T in [0.5, 5.0] in steps of $\Delta T = 0.5$. The best estimate of T is the one that yields the highest R_{et}^2 for all censuses provided that it is greater than 0.95.

Thus the model is able to fit with pretty decent accuracy the main biodiversity metrics used to characterize community structure such as RSA for the nine tropical forest plots studied (table 3.3 and figure 3.9) and SAR for the two plots for which this information was available, namely Barro Colorado and Pasoh (figure 3.10).

[6] It is important to prevent individual replacements of species having only one individual in order to constrain the CA configuration corresponding to the first census to the observed $S_e^{(1)}$ species.

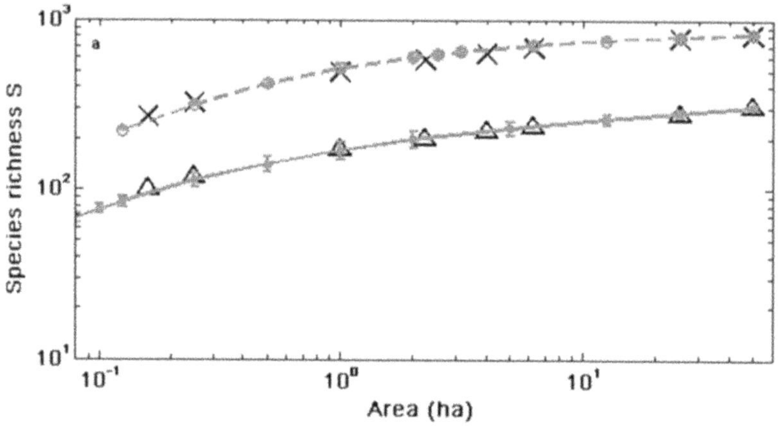

Figure 3.10. Species–area curves (SAR) for selected censuses of tropical forests. Predicted curves correspond to averages over 100 simulations for the best estimates of model parameters, and the error bars correspond to one std. Observed and predicted (grey line) number of tree species with dbh ⩾ 1 cm for sampling areas of different sizes at Barro Colorado (1990, triangles) and Pasoh (1987, crosses).

A central finding is the fact that the species niche width σ in all the nine tropical forests analyzed was roughly 0.08 (0.078±0.005) and hence that 95% of each species' niche corresponded to roughly 1/6 of the entire niche axis. It is quite remarkable that only a narrow set of niche widths allowed fitting the spatial and temporal dynamics in all tropical forests considered. This seemingly universal feature echoes the widespread convergence of functional traits leading to shade tolerance and nutritional and hydric niches axes leading to a large niche overlap in tropical tree species (Hubbell 2006). The fact that $2\sigma \approx 1/6$ implies by equation (3.32) that the strength of interspecific interactions is on average one-quarter of the intraspecific competition which, by construction, is equal to unity, i.e., $\langle \alpha_{ij} \rangle \approx 0.25 \langle \alpha_{ii} \rangle$. The ratio $(\langle \alpha_{ij} \rangle / \langle \alpha_{ii} \rangle) \approx 0.25$ corresponds to an intermediate value between the extreme claims of the *neutral theory of biodiversity* (Hubbell 2001) where species are functionally identical and have independent dynamics) and the classical niche-based model of community assembly (where interspecific competition is dominant).

3.4 Conclusion

The field of population biology focuses largely on single populations. But, except under controlled experimental situations, populations rarely live in isolation. Populations are typically embedded in communities, and their dynamics are strongly influenced by other members of the community. These feedbacks greatly complicate our understanding of the dynamics and present great challenges. Lotka and Volterra conceived a new formalism that envisages the units of a biological population as the statistical mechanics envisage molecules, atoms and electrons. That is, an approach that deals with such average effects as population density, population pressure, etc, in the same way in which thermodynamics deal with the average effects of gas concentration, gas pressures, etc. Since then, the Lotka–Volterra equations they proposed have been widely used to tackle problems not only in ecology but in other fields as diverse as chemistry, communication networks, economics, engineering, mathematics, laser physics or plasma physics.

In this chapter we exploited that the origin of these population equations is somehow rooted in statistical mechanics to further develop links between population/community dynamics with useful statistical mechanics tools and concepts. In the next chapter we will continue knitting connections by showing interesting analogies between ecology and evolution by one side with economics and finance for the other.

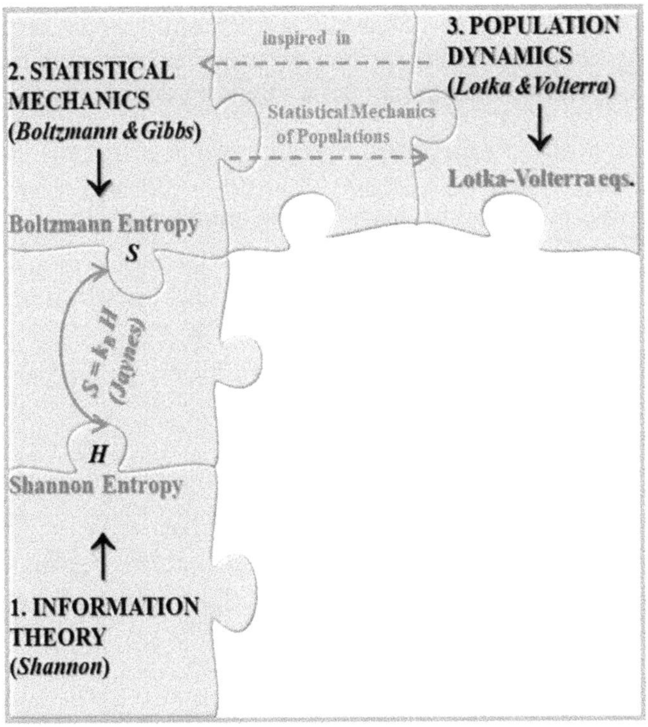

Appendix A: Equilibrium stability in population ecology

Stability theory addresses the stability of solutions of differential or difference equations used to model dynamical systems under perturbations of initial conditions.

A.1 Local and global stability

Let an ecological community of S species described by an autonomous continuous time model of the form:

$$\frac{dn_i}{dt} = n_i f_i(n_1, n_2, \ldots, n_S) \equiv n_i f_i(\mathbf{n}) \qquad i = 1, \ldots, S, \tag{A.1}$$

where $f_1(\mathbf{n}), f_2(\mathbf{n}), \ldots, f_S(\mathbf{n})$ are continuous functions in the positive orthant. We can denote the product $n_i f_i(\mathbf{n})$ as $F_i(\mathbf{n})$.

Similarly, if we use discrete time, and at time t let $N_i(t)$ denote the density of the ith species in an interaction among S species which we represent by a set of nonlinear difference equations:

$$n_i(t+1) = G_i(n_1, n_2, \ldots, n_S) \equiv G_i(\mathbf{n}) \qquad i = 1,\ldots, S, \tag{A.1'}$$

where, for convenience, we use n_i in place of $n_i(t)$; but in order to distinguish $n_i(t+1)$ from $n_i(t)$ we shall retain the argument of $n_i(t+1)$.

The simplest way to examine stability in a community model like equation (A.1) is by examining the eigenvalues of the so-called community matrix which is computed at an equilibrium of the model $\mathbf{n}^*$ that, by definition, verifies:

$$f_i(\mathbf{n}^*) = 0 \qquad i = 1,\ldots, S. \tag{A.2}$$

Similarly, for discrete time, we have that an equilibrium $\mathbf{n}^*$, by definition, must verify:

$$G_i(\mathbf{n}^*) = \mathbf{n}^* \qquad i = 1,\ldots, S. \tag{A.2'}$$

community matrix is the Jacobian matrix, given by:

$$J_{ij} \equiv \left.\frac{\partial\left(\frac{dn_i}{dt}\right)}{\partial n_j}\right|_{\mathbf{n}*} = \left.\frac{\partial F_i(\mathbf{n})}{\partial n_j}\right|_{\mathbf{n}*} = \left.\frac{\partial n_i}{\partial n_j} f_i(\mathbf{n})\right|_{\mathbf{n}*} + n_i^* \left.\frac{\partial f_i(\mathbf{n})}{\partial n_j}\right|_{\mathbf{n}*} = f_j(\mathbf{n}^*) + \left.\frac{\partial f_i(\mathbf{n})}{\partial n_j}\right|_{\mathbf{n}*} n_i^*, \tag{A.3}$$

and then, by equation (A.2), we can simply write the community or Jacobian matrix J_{ij} as:

$$J_{ij} = \left.\frac{\partial f_i(\mathbf{n})}{\partial n_j}\right|_{\mathbf{n}*} n_i^*. \tag{A.4}$$

However, the problem with this method is that it can only establish **local** stability or instability. On the other hand, ecosystems in the real world are subject to large perturbations of the initial state, and continual disturbances on the system dynamics that may produce important departures from equilibrium. Since the equations of population biology are nonlinear, their solutions, which can be represented as an S-dimensional surface, can give rise to quite complicated 'landscapes'. And therefore neighborhood stability analysis may give a misleading representation of the full global stability of the system.

If the dynamical equations are linear, local and global stability are identical. Unfortunately, while the linear approximation is very useful to approach many problems in physics, it rarely is a sensible approach in population biology. However, many biologically interesting models, although nonlinear, produce relatively simple landscapes, with one valley or hilltop whose sides slope ever upward or downward, respectively. In this case the local stability analysis correctly describes the global stability. Such circumstances are characterized by the existence of a *Lyapunov function* and constitute the basis of a powerful analytical method for establishing that an equilibrium is **globally** stable, i.e., stable relative to finite perturbations of the initial state. The so-called *direct* or *second* method of Lyapunov (LaSalle and Lefschetz 1961, Gurel and Lapidus, 1968, Willems, 1970, Strogatz 1994). There are many methods for constructing Lyapunov functions (Schultz 1965, Gurel and Lapidus 1968, Burton 1969, Hafstein 2007). However, unfortunately, there is no general way of knowing whether a Lyapunov function exists, let alone a straightforward procedure to construct it if it does exist.

In any event, for a given model it is possible to use computer simulations to investigate the behavior of the model for finite perturbations of its initial state. But computer simulations cannot guarantee that an equilibrium does indeed have a finite region of attraction. Certainly, this procedure becomes increasingly worse as the number of species in a given community increases.

In the next section of this appendix we will consider the local stability for two-dimensional systems. In the third and final section we will return to local and global stability, review the two methods outlined above, the one based on the eigenvalues of the Jacobian matrix and Lyapunov's method.

A.2 Stability for two-dimensional systems

Stability theory addresses the stability of solutions of differential equations and of trajectories of dynamical systems under small perturbations of initial conditions. We discuss here the stability of a general autonomous ordinary differential bi-dimensional system of equations of the form

$$\mathrm{d}x_1/\mathrm{d}t = f_1(x_1, x_2),\ \mathrm{d}x_2/\mathrm{d}t = f_2(x_1, x_2), \tag{A.5}$$

where f_1 and f_2 are given functions or maps. This system can be written more compactly in vector notation as

$$\mathrm{d}\mathbf{x}/\mathrm{d}t = \mathbf{f}(\mathbf{x}), \tag{A.5$'$}$$

where bold denotes a column vector with two entries:

$$\mathbf{x} = \begin{bmatrix} x_1 \\ x_2 \end{bmatrix}, \tag{A.6a}$$

and

$$\mathbf{f}(\mathbf{x}) = \begin{bmatrix} f_1(\mathbf{x}) \\ f_2(\mathbf{x}) \end{bmatrix}. \tag{A.6b}$$

Thus $\mathbf{x}$ represents a point in the phase plane, and $d\mathbf{x}/dt$ is the velocity vector at that point, which is given by the vector field or bi-dimensional map $\mathbf{f}(\mathbf{x})$. By flowing along the vector field, a phase point traces out a solution $\mathbf{x}(t)$, corresponding to a trajectory or *phase curve* winding through the phase plane (figure A.1).

However, what guarantee have we that the general nonlinear system $d\mathbf{x}/dt = \mathbf{f}(\mathbf{x})$ actually *has* solutions? Fortunately it turns out that there is an existence and uniqueness theorem for n-dimensional systems:

Existence and uniqueness theorem

Consider the initial value problem $d\mathbf{x}/dt = \mathbf{f}(\mathbf{x})$, $\mathbf{x}(0) = \mathbf{x}_0$. Suppose that $\mathbf{f}$ is *continuously differentiable*, i.e., $\mathbf{f}$ is continuous and that all its partial derivatives $\partial f_i/\partial x_j$, $i, j = 1,\ldots,n$, are continuous for $\mathbf{x}$ in some open connected set $\mathcal{D}$ contained in $\mathbf{R}^n$. Then for $\mathbf{x}_0$ in $\mathcal{D}$, the initial value problem has a solution $\mathbf{x}(t)$ on some time interval $(-\tau, \tau)$ about $t = 0$, and the solution is unique.

Fixed or singular points

Phase curves or phase trajectories in figure A.1 are solutions of

$$\frac{dx_1}{dx_2} = \frac{f_1(x_1, x_2)}{f_2(x_1, x_2)}. \tag{A.7}$$

We can imagine the entire phase plane as filled with such trajectories. In fact, through any point (x_1, x_2) there is a unique curve except at *fixed points* (x_1^*, x_2^*) where the vector field $\mathbf{f}(\mathbf{x})$ vanishes, i.e.,

$f_1(x_1^*, x_2^*) = f_2(x_1^*, x_2^*) = 0$.

This is why fixed points are also called *singular points.*

It turns out that for systems of nonlinear equations in general it is impossible to find the trajectories analytically. Even when explicit formulas are available, they are often too complicated to provide much insight. However, something we can do is to determine the *qualitative* behavior of the solutions. That is, to find the system's phase portrait directly from the properties of the vector field $\mathbf{f}(\mathbf{x})$. To do this we will use the ***linearization*** technique developed earlier for one-dimensional systems, namely a Taylor expansion around fixed points. The hope of this *linear stability analysis* is that we can approximate the phase portrait near a fixed point by that of a corresponding linear system, so that we can classify fixed points of *nonlinear* systems. More rigorously speaking, there is a theorem about the local behavior of

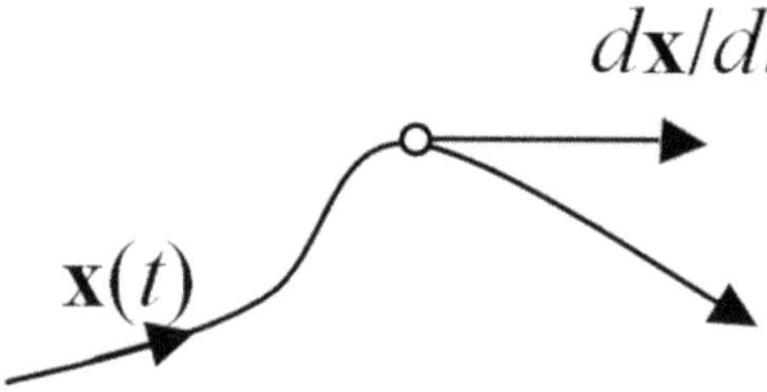

Figure A.1.

dynamical systems in the neighborhood of a certain type of equilibrium points which asserts that linearization is effective in predicting qualitative patterns of behavior.

Linear stability analysis around fixed points, the linearization theorem of Hartman–Grobman

Suppose the map **f** is *smooth*, i.e., it is at least differentiable everywhere (hence continuous) has an equilibrium state **x***: that is, **f(x*)** = 0. Then, the *Hartman–Grobman theorem* or *linearization theorem* states that the behavior of a dynamical system in a domain near a ***hyperbolic equilibrium point*** (we will define in a moment this kind of equilibrium) is qualitatively the same as the behavior of its linearization near this equilibrium point. Therefore, when dealing with such dynamical systems one can use the simpler linearization of the system to analyze its behavior around equilibria.

Just for simplicity of expression let us make the change of coordinates $x = x_1 - x_1^*$, $y = x_2 - x_2^*$, that moves the singular point to the origin $x = 0$ and $y = 0$. Then (0,0) is a singular point of the transformed equation (A.7′):

$$\frac{\mathrm{d}x}{\mathrm{d}y} = \frac{f_1(x, y)}{f_2(x, y)}. \tag{A.7′}$$

If f_1 and f_2 are analytic functions near (0,0), by definition of an analytic function, we can expand f_1 and f_2 in a Taylor series and, retaining only the linear terms, we get

$$\frac{\mathrm{d}x}{\mathrm{d}y} = \frac{f_{1x}x + f_{1y}y}{f_{2x}x + f_{2y}y}, \tag{A.8}$$

where the f_{ij} denote the partial derivative of the function f_i (i = 1 or 2) respect to the direction j = x or y evaluated at the origin, i.e., $f_{1x} = \left.\frac{\partial f_1}{\partial x}\right|_{(0,0)}$, $f_{1y} = \left.\frac{\partial f_1}{\partial y}\right|_{(0,0)}$, $f_{2x} = \left.\frac{\partial f_2}{\partial x}\right|_{(0,0)}$, $f_{2y} = \left.\frac{\partial f_2}{\partial y}\right|_{(0,0)}$. These four numbers define the *Jacobian* matrix **J**:

$$\mathbf{J} = \begin{pmatrix} f_{1x} & f_{1y} \\ f_{2x} & f_{2y} \end{pmatrix} \equiv \begin{pmatrix} \frac{\partial f_1}{\partial x} & \frac{\partial f_1}{\partial y} \\ \frac{\partial f_2}{\partial x} & \frac{\partial f_2}{\partial y} \end{pmatrix}_{(0,0)}. \tag{A.9}$$

Therefore, equation (A.5′) is equivalent, to first order (i.e., linear approximation) to:

$$\frac{d\mathbf{x}}{dt} = \mathbf{J}\mathbf{x}. \tag{A.10}$$

Let λ_1 and λ_2 be the eigenvalues of **J**; given by equating the determinant of $\mathbf{J} - \lambda\,\mathbf{I}$ (where **I** is the identity matrix):

$$\begin{vmatrix} f_{1x} - \lambda & f_{1y} \\ f_{2x} & f_{2y} - \lambda \end{vmatrix} = 0, \tag{A.11}$$

i.e., λ_1 and λ_2 are the roots of the second order ***characteristic polynomial***:

$$\lambda^2 - (f_{1x} + f_{2y})\lambda + f_{1x}f_{2y} - f_{1y}f_{2x} = 0, \tag{A.12}$$

which can be re-written as:

$$\lambda^2 - \mathrm{Tr}\mathbf{J}\lambda + \det \mathbf{J} = 0, \tag{A.13}$$

where 'Tr' denotes the trace of matrix **J** (the sum of diagonal elements) and 'det' its determinant $|\mathbf{J}|$. Therefore, we get

$$\begin{aligned} \lambda_1 &= 1/2\left(\mathrm{Tr}\mathbf{J} + \sqrt{\mathrm{Tr}\mathbf{J}^2 - 4\,\mathrm{det}\mathbf{J}}\right), \\ \lambda_2 &= 1/2\left(\mathrm{Tr}\mathbf{J} - \sqrt{\mathrm{Tr}\mathbf{J}^2 - 4\,\mathrm{det}\mathbf{J}}\right). \end{aligned} \tag{A.14}$$

In general, λ_1 and λ_2 are complex numbers. An equilibrium $\mathbf{x}^*$ is hyperbolic if no eigenvalue of the linearization has real part equal to zero, i.e., hyperbolic equilibrium implies that $\mathrm{Re}(\lambda_1) \neq 0$ and $\mathrm{Re}(\lambda_2) \neq 0$.

The typical situation is for the eigenvalues to be distinct: $\lambda_1 \neq \lambda_2$. In this case, a theorem of linear algebra states that the corresponding eigenvectors $\mathbf{v}_1$ and $\mathbf{v}_2$ of **J** are linearly independent, and hence span the entire plane. In particular, any initial condition $\mathbf{x}_0$ can be written as a linear combination of eigenvectors, say

$$\mathbf{x}_0 = c_1\mathbf{v}_1 + c_2\mathbf{v}_2, \tag{A.15}$$

where c_1 and c_2 are arbitrary constants and the eigenvector $\mathbf{v}_i$ associated with the eigenvalue λ_i given by

$$\mathbf{v}_i = \left(1 + p_i^2\right)^{-1/2}\begin{bmatrix} 1 \\ p_i \end{bmatrix}, \quad p_i = \frac{\lambda_i - f_{1x}}{f_{1y}}, \quad f_{2x} \neq 0, \quad i = 1,2. \tag{A.16}$$

This allows us to write down the general solutions of (A.10) simply as

$$\mathbf{x} = c_1\mathbf{v}_1 e^{\lambda_1 t} + c_2\mathbf{v}_2 e^{\lambda_2 t}. \tag{A.17}$$

This is a general solution because it is a linear combination of solutions to equation (A.10), and hence is itself a solution. In addition, it satisfies the initial condition $\mathbf{x}(0) = \mathbf{x}_0$, and so by the existence and uniqueness theorem, it is the *only* solution.

If the eigenvalues are equal, i.e., equation (A.13) has a double root $\lambda_1 = \lambda_2 = \lambda$, the solutions are proportional to $(c_1 + c_2 t)\exp[\lambda t]$.

The mathematician Henri Poincaré distinguished four different singular points of differential equations. These are the ***node***, the ***saddle***, the ***focus*** and the ***center***.

Figure A.2 summarizes the possibilities in the so-called ***Poincaré diagram***, i.e., the (tr **A**, det **A**) parameter plane, which includes the parable $\Delta \equiv 1/4 \text{ tr } \mathbf{A}^2 - \det \mathbf{A} = 0$.

If det **A** < 0, then λ_1 and λ_2 are real and of opposite signs, regardless of the sign of tr **A**.

Usually, solutions go to infinity as $t \to \infty$ so this case is considered to be unstable. Figure A.2 shows the appearance of some trajectories near this kind of fixed point, denoted a **saddle point**. This type of behavior is found in the region below the horizontal axis of the (tr **A**, det **A**) parameter plane shown in the summary of figure A.2.

If det **A** > 0, then any of the following can happen:

(A) det **A** < 1/4 tr $\mathbf{A}^2$ (i.e., below the parable): In this case λ_1 and λ_2 are real. We then have two possibilities:

II-(A).1. If tr **A** < 0: In this case $\lambda_1 < 0$ and $\lambda_2 < 0$. Solutions are both decreasing exponentials so that the *fixed point is stable*, denoted a **stable node** or **sink** (located in the Poincaré diagram between the horizontal axis and the parable, to the left hand side of figure A.2).

II-(A).2. If tr **A** > 0: In this case $\lambda_1 > 0$ and $\lambda_2 > 0$. Solutions are both increasing exponentials so that the *fixed point is unstable*, denoted an **unstable node** or **source** (between the horizontal axis and the parable, to the rhs of figure A.2).

(B) det **A** > 1/4 tr $\mathbf{A}^2$ (i.e., above the parable): In this case λ_1 and λ_2 are complex.

We then have three possibilities:

II-(B).1. If tr **A** < 0: In this case $\text{Re}(\lambda_1) < 0$ and $\text{Re}(\lambda_2) < 0$. Solutions are oscillations with decreasing amplitude so that the *fixed point* is a **stable focus** or **spiral**

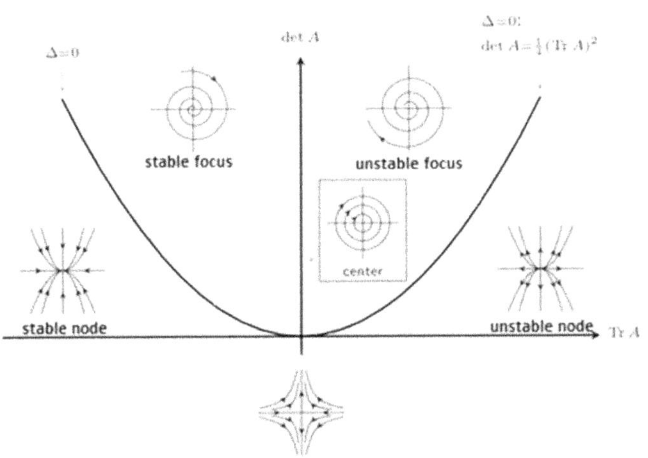

Figure A.2. Poincaré diagram. Classification of phase portraits in the (Tr **A**, det **A**)-plane. Author Freesodas. Source: Gimp.

sink (located in the Poincaré diagram between the vertical axis and the parable, to the lhs of figure A.2).

II-(B).2. If tr **A** > 0: In this case $\mathrm{Re}(\lambda_1) > 0$ and $\mathrm{Re}(\lambda_2) > 0$. Solutions are oscillations with increasing amplitude so that the *fixed point* is an **unstable focus** or **spiral source** (located in the Poincaré diagram between the vertical axis and the parable, to the rhs of figure A.2).

II-(B).3. If tr **A** = 0: In this case $\mathrm{Re}(\lambda_1) = \mathrm{Re}(\lambda_2) = 0$. Solutions are periodic, with constant amplitude, and thus the phase curves are ellipses. This corresponds to a **center**, but is a marginal case. Centers are not stable in the usual sense, they are *neutrally stable*; a small perturbation from one phase curve does not die out in the sense of returning to the original unperturbed curve. The perturbation simply gives another solution. This implies that, in general, nonlinear terms will either stabilize or destabilize the system. In the case of center singularities, determined by the linear approximation to $f_1(x, y)$ and $f_2(x, y)$, we must look at the higher order (rather than linear) terms to determine whether or not it is really a spiral and hence whether it is stable or unstable.

Summary

- Fixed points are **stable** when **the real part of λ_1 and λ_2 are negative**.
- There are four types of fixed points:

1. A **node** if λ_1 and λ_2 are real, non-null, with the same sign; if both λ_1 and λ_2 are negative (positive) the node is **stable** (**unstable**).
2. A **saddle point** if the signs of λ_1 and λ_2 are opposite.
3. A **focus** when λ_1 and λ_2 are complex (with real part different from 0). Negative real parts for λ_1 and λ_2 imply a **stable focus**, whereas positive real parts for λ_1 and λ_2 mean an **unstable focus**.
4. A **center** if λ_1 and λ_2 are purely imaginary the fixed point is. In this case we have to go beyond the linear stability analysis and look at the nonlinear terms to determine whether or not it is really a spiral and hence whether it is stable or unstable.

Limit cycles and Kolmogorov's theorem for predator–prey systems

An interesting question about trajectories that spiral outward from an unstable equilibrium is: do they spiral outward without bound until they intersect one of the axes and one of the species goes extinct? Or do they settle on a particular orbit which is itself stable? Such orbits are called stable **limit cycles**. A limit cycle is a closed trajectory such that neighboring trajectories are not closed; they spiral either toward (stable limit cycle) or away from the limit cycle (unstable limit cycle).

To elucidate the question of the fate of unstable spirals there are both negative theorems, which rule out closed orbit solutions in the phase plane, as well as the Poincaré–Bendixon theorem which establishes that closed orbits exist under particular conditions. Before introducing these theorems, at the end of the next section, we will present a theorem by the Russian mathematician Andrei Kolmogorov for bi-dimensional predator–prey systems.

Kolmogorov's theorem

Given a bi-dimensional system like equation (A.5), if

$$\begin{aligned} f_1(x_1, x_2) &= x_1 f(x_1, x_2), \\ f_2(x_1, x_2) &= x_2 g(x_1, x_2), \end{aligned} \tag{A.18}$$

where $f(x_1, x_2)$ and $g(x_1, x_2)$ can be interpreted as the per capita growth rates for each species, provided that

(i) the functions f and g are continuous and differentiable in the domain $x_1 > 0$ and $x_2 > 0$

$$\begin{aligned}
&\textbf{(ii)} \quad \frac{\partial f}{\partial x_2} < 0 \\
&\textbf{(iii)} \quad \frac{\partial f}{\partial x_1} x_1 + \frac{\partial f}{d x_2} x_2 < 0 \\
&\textbf{(iv)} \quad \frac{\partial g}{\partial x_2} \leqslant 0 \\
&\textbf{(v)} \quad \frac{\partial g}{\partial x_1} x_1 + \frac{\partial g}{d x_2} x_2 > 0 \\
&\textbf{(vi)} \quad f(0,0) > 0 \\
&\textbf{(vii)} \quad f(0, A) = 0 \\
&\textbf{(viii)} \quad f(B, 0) = 0 \\
&\textbf{(ix)} \quad f(C, 0) = 0 \\
&\textbf{(x)} \quad B > C
\end{aligned} \tag{A.19}$$

where A, B and C are three positive quantities, then this system has either a stable point of equilibrium or a stable limit cycle.

A.3 Some general theorems[7]

We will introduce some valuable theorems that we will accept without proving them (for proofs of these theorems see Goh (1980)).

Local stability: the real parts of the eigenvalues of the Jacobian matrix must be negative

[7] This section is mainly based on chapters 1, 3 and 5 of the thorough study on stability by Goh (1980) and chapter 2 of May (1974).

Suppose the autonomous system (A.1) has a positive equilibrium at $\mathbf{n}^*$ and let $x_i = n_i - n_i^*$ for $i = 1, 2, \ldots, S$ denote a small departure of each species density from its equilibrium value. Performing a first order Taylor expansion around the equilibrium we get:

$$n_i f_i(\mathbf{n}) = n_i^* \left(f_i(\mathbf{n}^*) + \sum_{j=1}^{S} \frac{\partial f_i(\mathbf{n})}{\partial n_j}\bigg|_{\mathbf{n}*} x_j + O(x^2) \right) = n_i \\ * \left(0 + \sum_{j=1}^{S} \frac{\partial f_i(\mathbf{n})}{\partial n_j}\bigg|_{\mathbf{n}*} x_j + O(x^2) \right). \qquad \text{(A.20)}$$

Therefore, substituting (A.20) into (A.1) and using (A.4), the linearized dynamics is given by

$$\frac{\mathrm{d}x_i}{\mathrm{d}t} = \sum_{j=1}^{S} n_i^* J_{ij} x_j \qquad i = 1, \ldots, S. \qquad \text{(A.21)}$$

It turns out that we have this valuable theorem for continuous time models:

Theorem 1. The equilibrium $\mathbf{n}^* = (n_1^*, n_2^*, \ldots, n_S^*)$ of an autonomous continuum time system is locally stable if all the real parts of the eigenvalues of the Jacobian matrix $J_{ij} \equiv \frac{\partial f_i(\mathbf{n})}{\partial n_j}\Big|_{\mathbf{n}*} n_i^*$ are negative.

Thus this theorem generalizes the two stability analysis for bi-dimensional systems of the previous section.

In the case of a Lotka–Volterra generalized linear model we have:

$$\frac{\mathrm{d}n_i}{\mathrm{d}t} = r_i n_i \left(1 + \sum_{j=1}^{S} I_{ij} n_j \right) \qquad i = 1, \ldots, S. \qquad \text{(A.22)}$$

And thus we have,

$$J_{ij} = r_i I_{ij} n_i^* \qquad \text{(A.23)}$$

The equilibrium $\mathbf{n}^*$ is locally stable if all the real parts of the eigenvalues of the matrix $[r_i a_{ij} n_i^*]$ are negative. Note that in general the stability properties of the matrix $[r_i a_{ij} n_i^*]$ are different from those of the matrix $\mathbf{I} = [I_{ij}]$.

For the discrete time description (A.1′), to first order, we have:

$$x_i(t+1) = \sum_{j=1}^{S} \left.\frac{\partial G_i(\mathbf{n})}{\partial n_j}\right|_{\mathbf{n}*} x_j. \qquad \text{(A.21′)}$$

And, a theorem similar to theorem 1 but for discrete time models (Goh 1980) is:

> **Theorem 1 ′.** The equilibrium $\mathbf{n}^* = (n_1^*, n_2^*, \ldots, n_S^*)$ of a set of discrete difference equations is locally stable if the modulus of all the eigenvalues of the matrix $\left.\frac{\partial G(\mathbf{n})}{\partial n_j}\right|_{\mathbf{n}*}$ are less than one.

Global stability: Lyapunov functions

There exists a method to determine whether a system is globally stable. It involves finding a function known as a *Lyapunov function.* The problem is that the existence of a Lyapunov function is often difficult to determine for multispecies models and, consequently, this approach has a limited utility. The discussion will be facilitated by considering physical systems analogous to the biological ones.

Consider an autonomous system of differential equations

$$\mathrm{d}x_i/\mathrm{d}t = f_i(\mathbf{x}) \qquad \text{(A.24)}$$

with a fixed point at $\mathbf{x}^* = (x_1^*, x_2^*, \ldots, x_S^*)$.

Definition: A *Lyapunov function*, for this system is a continuously differentiable, real valued function $V(\mathbf{x})$ with the following properties:

i. $V(\mathbf{x}) > 0$ for all $\mathbf{x} \neq \mathbf{x}^*$, and $V(\mathbf{x}^*) = 0$. (We say that V is *positive definite.*)

ii. $\frac{\mathrm{d}V(x)}{\mathrm{d}t} = \sum_{i=1}^{S} \frac{\partial V}{\partial x_i}\frac{\mathrm{d}x_i}{\mathrm{d}t} = \sum_{i=1}^{S} \frac{\partial V}{\partial x_i} f_i(\mathbf{x}) < 0$ for all $\mathbf{x} \neq \mathbf{x}^*$. (All trajectories flow 'downhill' toward $\mathbf{x}^*$.)

> **Theorem 2.** The equilibrium $\mathbf{x}^*$ of an autonomous continuum time system is locally asymptotically stable, i.e., for all initial conditions $\mathbf{x}(t) \to \mathbf{x}^*$, as $t \to \infty$, if there exists a Lyapunov function for $\mathbf{x}^*$.

Intuitively, under conditions (**i.** and **ii.**) all trajectories move monotonically down the graph of $V(\mathbf{x})$ toward $\mathbf{x}^*$ (figure A.3).

The solutions can't get stuck anywhere else because if they did, V would stop changing, but by assumption, $\mathrm{d}V/\mathrm{d}t < 0$ everywhere except at $\mathbf{x}^*$.

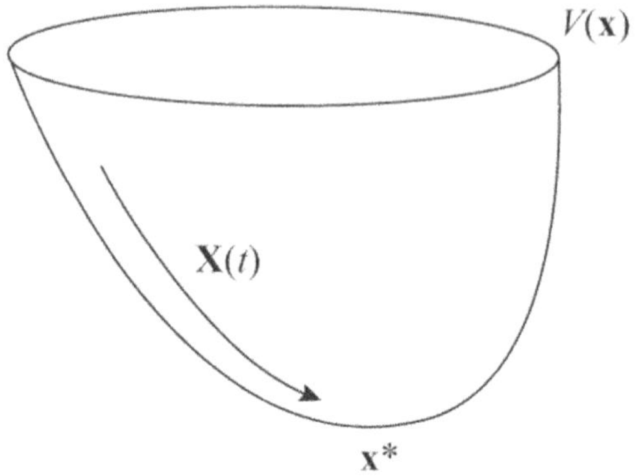

Figure A.3. Figure I3 of my IOP book *Ecological Modelling and Ecophysics.* Reproduced from Fort (2020).

For physical systems the direct method of Lyapunov generalizes the principle that a system, which continuously dissipates energy until it attains an equilibrium, is stable. If a physical system, for example a vibrating spring and mass, dissipates energy over time and the energy is never restored then eventually the system must reach some final resting state. This final state is called the *attractor*. However, finding a function that gives the precise energy of a physical system can be difficult, and for biological systems, the concept of energy may not be applicable. Lyapunov's realization was that stability can be proven without requiring knowledge of the true physical energy, provided a Lyapunov function can be found to satisfy the above constraints.

Goh (1977) has discussed a Lyapunov function that fits all Lotka–Volterra models:

$$V(\mathbf{n}) = \sum_{i=1}^{S} \mathrm{k}_i (n_i - n_i^* - n_i \ln(n_i/n_i^*)), \tag{A.25}$$

where the k_i are constants. If the k_i exist such that $\mathrm{d}V/\mathrm{d}t$ is always negative except at $\mathbf{n}^*$ (where it is zero) then the system is globally stable.

The problem is how to obtain the k_i so that they satisfy condition $\mathrm{d}V/\mathrm{d}t < 0$. For simple examples this is easy, but for more complicated examples it is not

Ruling out closed orbits

Closed orbits can be ruled out for the following systems[8]:

(A) ***Gradient systems***

That is, suppose the system can be written in the form $\mathrm{d}\mathbf{x}/\mathrm{d}t = -\nabla V$, for some continuously differentiable, single-valued scalar function $V(\mathbf{x})$ (this vector equality is written for each coordinate i as $\mathrm{d}x_i/\mathrm{d}t = -\partial V/\partial x_i$). Such a system is called a *gradient* with *potential function V*.

[8] For proofs we refer the reader to section 7.2 of Strogatz (1994).

(B) Systems with a Lyapunov function

If a Lyapunov function exists, then closed orbits are forbidden

Poincaré–Bendixon theorem

The Poincaré–Bendixon theorem is the main tool which historically has been used to show that a dynamical system has a stable cycle limit.

Poincaré–Bendixon theorem. Suppose that: (i) R is a closed, bounded subset of the plane; (ii) $d\mathbf{x}/dt = \mathbf{f}(\mathbf{x})$ is a continuously differentiable vector field on an open set containing R; (iii) R does not contain any fixed points; (iv) There exists a trajectory C that is 'confined' in R, this means that it starts in R and stays in R for all future time.Then either C is a closed orbit, or it spirals toward a closed orbit as $t \to \infty$. Thus, in either case, R ***contains a closed orbit.***

For a proof of this theorem we refer the interested reader to Coddington and Levinson (1955) or Wiggins (1990).

Figures:

(*): https://www.britannica.com/story/thomas-malthuss-250th-birthday

(**): https://www.researchgate.net/publication/292314245_Modeling_single_species_populations_with_MATLAB/figures?lo=1

References

Abrams P A 1983 Arguments in favor of higher order interactions *Am. Nat.* **121** 887–91

Adeel Z, Safriel U, Niemeijer D and White R 2005 Ecosystems and human well-being: desertification synthesis *A Report of the Millennium Ecosystem Assessment, World Resources Institute, Washington DC, USA*

Alexeyev V and Levich A 1997 A search for maximum species abundances in ecological communities under conditional diversity optimization *Bull. Math. Biol.* **59** 649–77

Arbiza J, Mirazo S and Fort H 2010 Viral quasispecies profiles as the result of the interplay of competition and cooperation *BMC Evol. Biol.* **10** 137

Banavar J R, Maritan A and Volkov I 2010 Applications of the principle of maximum entropy: from physics to ecology *J. Phys. Condens. Matter* **22** 063101

Brown J H, Whitham T G, Ernest S K M and Gehring C A 2001 Complex species interactions and the dynamics of ecological systems: long-term experiments *Science* **293** 643–50

Burton T A 1969 On the construction of Lyapunov functions *SIAM J. Appl. Math.* **17** 1078–85

Carpenter S 2001 Alternate states of ecosystems: evidence and some implications *Ecology: Achievement and Challenge* ed N Huntly, M C Press and S Levin (Oxford: Blackwell) pp 357–83

Coddington E A and Levinson N 1955 The Poincaré–Bendixon theory of two-dimensional autonomous systems *Theory of Ordinary Differential Equations* (New York: McGraw-Hill) pp 389–403

Dregne H E 1986 Desertification of arid lands *Physics of Desertification* ed F El-Baz and M H A Hassan (Dordrecht: Martinus Nijhoff)

Durrett R and Levin S 1998 Spatial aspects of interspecific competition *Theor. Pop. Biol.* **53** 30–43

ForestGEO 2021 https://forestgeo.si.edu/ (accessed 07-9-2021)

Fort H 2020 *Ecological Modelling and Ecophysics: Agricultural and Environmental Applications* (Bristol: IOP Publishing) see Exercise 2.5

Fort H 2018 On predicting species yields in multispecies communities: quantifying the accuracy of the linear Lotka–Volterra generalized model *Ecol. Modell.* **387** 154–62

Fort H 2013 Statistical mechanics ideas and techniques applied to selected problems in ecology *Entropy* **15** 5237–76

Glauber R J 1963 Time-dependent statistics of the Ising model *J. Math. Phys.* **4** 294–307

Hamann J R and Bianchi L M 1970 Stochastic population mechanics in the relational systems formalism: Volterra–Lotka ecological dynamics *J. Theor. Biol.* **28** 175–84

Harte J *et al* 2008 Maximum entropy and the state-variable approach to macroecology *Ecology* **89** 2700–11

Harte J 2011 *Maximum Entropy and Ecology: A Theory of Abundance, Sistribution, and Energetics* (Oxford: Oxford University Press)

He F 2010 Maximum entropy, logistic regression, and species abundance *Oikos* **119** 578–82

Goel N S, Maitra S C and Montroll E W 1971 On the Volterra and other nonlinear models of interacting populations *Rev. Mod. Phys.* **43** 231–76

Goh B S 1977 Global stability in many species systems *Am. Nat.* **111** 135–43

Goh B S 1980 *Management and Analysis of Biological Populations* (Amsterdam: Elsevier)

Gurel O and Lapidus L 1968 Stability via Lyapunov's second method *Ind. Eng. Chem.* **60** 1–26

Hafstein S F 2007 An algorithm for constructing Lyapunov functions *Electronic Journal of Differential Equations Monograph 08* http://ejde.math.txstate.edu or http://ejde.math.unt.edu

Halty V *et al* 2017 Modelling plant interspecific interactions from experiments of perennial crop mixtures to predict optimal combinations *Ecol. Appl.* **27** 2277–89

Holmgren M, Scheffer M and Huston M A 1997 The interplay of facilitation and competition in plant communities *Ecology* **78** 1966–75

Hubbell S P 2006 Neutral theory and the evolution of ecological equivalence *Ecology* **87** 1387–98

Hubbell S P 2001 *The Unified Neutral Theory of Biodiversity and Biogeography* (Princeton, NJ: Princeton University Press)

Hubbell S P, Ahumada J A, Condit R and Foster R B 2001 Local neighborhood effects on long-term survival of individual trees in a neotropical forest *Ecol. Res.* **16** S45–61

Hutchinson G E 1957 Concluding remarks – *Cold Spring Harbor Symp. on Quantitative Biology* **vol 22** 415–27

Kerner E H 1957 A statistical mechanics of interacting biological species *Bull. Math. Biophys* **19** 121–46

Kerner E H 1959 Further considerations on the statistical mechanics of biological associations *Bull. Math. Biophys* **21** 217–55

Kerner E H 1972 *Gibbs Ensemble: Biological Ensemble. The Application of Statistical Mechanics to Ecological, Neural and Biological Networks* (New York: Gordon and Breach)

Kingsland S 2015 Alfred J. Lotka and the origins of theoretical population ecology *Proc. Natl Acad. Sci.* **112** 9493–5

LaSalle J and Lefschetz S 1961 *Stability by Lyapunov's Direct Method* (New York: Academic)

Legates D R and McCabe G J 1999 Evaluating the use of 'goodness-of-fit' measures in hydrologic and hydroclimatic model validation *Water Resour. Res.* **35** 233–41

Levins R 1968 *Evolution in Changing Environments* (Princeton, NJ: Princeton University Press)

Li J 2017 Assessing the accuracy of predictive models for numerical data: not r nor r^2, why not? Then what? *PLoS One* **12** e0183250

Li J and Heap A 2008 A Review of Spatial Interpolation Methods for Environmental Scientists. Geoscience Australia, Record 2008/23, 137 pp

Lotka A J 1925 *Elements of Physical Biology* (Baltimore, MD: Williams & Wilkins)

Lotka A J 1922a Natural selection as a physical principle *Proc. Natl Acad. Sci.* **8** 151–4

Lotka A J 1922b Contribution to the energetics of evolution *Proc. Natl Acad. Sci.* **8** 147–51

MacArthur R H and Levins R 1967 The limiting similarity, convergence and divergence of coexisting species *Am. Nat.* **101** 377–85

May R M 1974 *Stability and Complexity in Model Ecosystems* (Princeton, NJ: Princeton University Press)

May R M 1976 Simple mathematical models with very complicated dynamics *Nature* **261** 459–67

Morin P J 2011 *Community Ecology* (Chichester: Wiley)

Nash J E and Sutcliffe J V 1970 River flow forecasting through conceptual models, I: a discussion of principles *J. Hydrol.* **10** 282–90

Nee S and Colegrave N 2006 Ecology: paradox of the clumps *Nature* **441** 417–8

Novak M *et al* 2016 Characterizing species interactions to understand press perturbations: what is the community matrix? *Annu. Rev. Ecol. Evol. Syst.* **47** 409–32

Noy-Meir I 1975 Stability of grazing systems: an application of predator–prey graphs *J. Ecol.* **63** 459–81

Odum H T and Pinkerton R C 1955 Time's speed regulator: the optimum efficiency for maximum power output in physical and biological systems *Am. Sci.* **43** 331–43

Pueyo S, He F and Zillio T 2007 The maximum entropy formalism and the idiosyncratic theory of biodiversity *Ecol. Lett.* **10** 1017–28

Rees M P *et al* 1996 Quantifying the impact of competition and spatial heterogeneity on the structure and dynamics of a four-species guild of winter annuals *Am. Nat.* **147** 1–32

Roxburgh S H and Wilson J B 2000 Stability and coexistence in a lawn community: mathematical prediction of stability using a community matrix with parameters derived from competition experiments *Oikos* **88** 395–408

Schultz D G 1965 The generation of Lyapunov functions *Advances in Control Systems* ed C T Leondes vol 2 (New York: Academic) pp 1–64

Sharkovsky A N, Maistrenko Y L and Romanenko E Y 1993 *Difference Equations and Their Applications* (Dordrecht: Springer)

Shipley B, Vile D and Garnier E 2006 From plant traits to plant communities: a statistical mechanistic approach to biodiversity *Science* **314** 812–4

Silvertown J, Holtier S, Johnson J and Dale P 1992 Cellular automaton models of interspecific competition for space: the effect of pattern on process *J. Ecol.* **80** 527–33

Strogatz S H 1994 *Nonlinear Dynamics and Chaos. With Applications to Physics, Biology, Chemistry, and Engineering* (Reading, MA: Perseus Books)

Turner P E and Chao L 1999 Prisoner's dilemma in an RNA virus –3 *Nature* **398** 441–2

Uriarte M, Condit R, Canham C and Hubbell S 2004 A spatially explicit model of sapling growth in a tropical forest: does the identity of neighbours matter? *J. Ecol.* **92** 348–60

Vandermeer J H 1969 The competitive structure of communities: an experimental approach with protozoa *Ecology* **50** 362–71

Vandermeer J H and Goldberg D E 2013 *Population Ecology: First Principles* 2nd edn (Princeton, NJ: Princeton University Press)

van der Vaart H R 1973 A comparative investigation of certain difference equations and related differential equations *Bull. Math. Biol.* **35** 195–211

Verhulst P F 1838 Notice sur la loi que la population poursuit dans son accroissement *Correspondance Mathématique et Physique* **10** 113–21

Volkov I, Banavar J R, Hubbell S P and Maritan A 2009 Inferring species interactions in tropical forests *Proc. Natl Acad. Sci.* **106** 13854–9

Volterra V 1931 *Leçons sur la Théorie Mathématique de la Lutte pour la Vie* (Paris: Gauthier-Villars)

Volterra V 1926 Fluctuations in the abundance of a species considered mathematically *Nature* **118** 558–60

Weissmann H and Shnerb N M 2016 Predicting catastrophic shifts *J. Theor. Biol.* **397** 128–32

Wiens J A 1984 On understanding a non-equilibrium world: myth and reality in community patterns and processes *Ecological Communities: Conceptual Issues and the Evidence* (Princeton, NJ: Princeton University Press) pp 439–57

Wiesmeier M 2015 Environmental indicators of dryland degradation and desertification *Environmental Indicators* ed R H Armon and O Hänninen (Berlin: Springer) ch 14

Wiggins S 1990 *Introduction to Applied Nonlinear Dynamical Systems and Chaos* (New York: Springer)

Wilcox B P, Rawls W J, Brakensiek D L and Wight J R 1990 Predicting runoff from rangeland catchments A: comparison of two models *Water Resour. Res.* **26** 2401–10

Willems J L 1970 *Stability Theory of Dynamical Systems* (London: Nelson)

Willmott C J 1984 On the evaluation of model performance in physical geography *Spatial Statistics and Models* ed G L Gaile and C J Willmott (Norwell, MA: D. Reidel) pp 443–60

Willmott C J 1981 On the validation of models *Phys. Geogr.* **2** 184–94

Wolfram S 2002 *A New Kind of Science* (Champaign, IL: Wolfram Media)

Wolfram S 1994 *Cellular Automata and Complexity, Collected Papers* (Reading, MA: Addison-Wesley)

Wilson S D and Keddy P A 1986 Species competitive ability and position along a natural stress/disturbance gradient *Ecology* **67** 1236–42

IOP Publishing

Forecasting with Maximum Entropy
The interface between physics, biology, economics and information theory
Hugo Fort

Chapter 4

Economics as physics, economics as biology

'Nevertheless, their determination will render Economics a science as exact as many of the physical sciences.'

—W Stanley Jevons (1871, *The Theory of Political Economics*)

'The essential point to grasp is that in dealing with capitalism we are dealing with an evolutionary process.'

—Joseph Schumpeter (1942, *Capitalism, Socialism, and Democracy*)

We start by discussing the intimate connection between the history of physics and the history of economics as well as the remarkable mutual influence between both disciplines throughout the 19th century. In particular, we review the tradition of formulating **economics as physics** to become a *positive* science. We also comment on the other side of the coin: how economics provided J J Joule with many of the metaphorical and experimental guidelines he needed to translate thermodynamic phenomena into quantifiable terms.

In section 4.2, we present the basics of the neoclassical theory, inspired in the mechanics metaphor, which forms the basis of mainstream economics. Thus we discuss the central concept of marginal utility that replaced the labor theory of value and other older ideas of classical economics. Because economics is essentially the science of how people use and value economic goods in order to achieve their limitless wants and needs with limited and scarce resources at hand, marginal thinking is ubiquitous in all areas of economics. We also analyze how this marginal point of view can be used to derive a general equilibrium theory.

In section 4.3 first we consider the criticism to mainstream economics. One of the most important objections is that it relies almost exclusively on the axiomatic-deductive approach while empirical evidence contradicting neoclassical assumptions is generally ignored. Another critique is about rational choice theory which assumes that people always make optimal decisions that provide them with the greatest benefit and satisfaction. This would happen in an ideal world. However, we don't

doi:10.1088/978-0-7503-3931-5ch4

live in a perfect world; in reality, people are often moved by emotions and external factors. We then move to **economics as biology** by introducing the main ideas of evolutionary economics. Interestingly, evolutionary ideas have been useful to explain the nature of the processes of innovation and the institutions supporting them, thus showing that an evolutionary theoretical perspective can provide useful heuristics for applied research.

Sections 4.4–4.6 discuss different aspects of evolution by natural selection in biology. In section 4.4 we show how to model mathematically the selection process of competing populations. We first consider the simplest kind of selection of population types that have fixed fitnesses and then the case of frequency dependent selection, in the form of the replicator dynamics equation (RDE). Section 4.5 is devoted to proving the equivalence of the RDE with the ecological generalized Lotka–Volterra equations of chapter 3. In section 4.6 we address the other engine of evolution, mutations. So we present different equations that extend those of pure selection, introduced in section 4.5.

In section 4.7 we analyze how the evolutionary concepts and models can be applied in economics. An important part of this program was carried out by economists Thorstein Veblen and Joseph Schumpeter.

We conclude this chapter by stating in section 4.8 the so-called 'Marshall's problem', namely finding a way to integrate physics and biology into economics in a way that both contains its orthodox core as well as provides a transdisciplinary perspective on economic complexity. This is a main theme of this book.

4.1 Economics as social physics, physics as Nature's economics[1]

The history of physics and the history of economics are intimately connected and the mutual influence between both disciplines was remarkable in the course of the 19th century. An interesting example of such confluence of both disciplines is *social physics* or *sociophysics*, a field of science which uses mathematical tools inspired by physics to understand the behavior of human crowds. Its roots can be traced back to the 17th century and it is closely related to the modern interdisciplinary research field of *econophysics* (Mantegna and Stanley 2000), which uses physics methods to describe economics.

Let us point to some landmarks in the development of social physics.

- The English philosopher Thomas Hobbes seems to have been the first to envisage the concept of social physics. In 1636, during a trip to Florence, Italy, Hobbes met the famous astronomer and physicist Galileo Galilei, who was well-known for his ideas about the motion of bodies. Hobbes was greatly influenced by Galileo and he began to outline the idea of representing the 'physical phenomena' of society in terms of the laws of motion. Although there was no explicit mention of 'social physics', the program of examining society with scientific methods began here.

[1] The title of this section is borrowed from Philip Mirowski's (1989) book, this volume inspired the content of this section.

- Later on, the French social thinker Henri de Saint-Simon introduced the idea of describing society using laws similar to those of the physical and biological sciences (1803). His student and collaborator was Auguste Comte, a French philosopher regarded as the founder of sociology, who coined the term in an essay appearing in *Le Producteur*, a journal project by Saint-Simon (Iggers 1959).
- In Comte's words: 'Social physics is that science which occupies itself with social phenomena, considered in the same light as astronomical, physical, chemical, and physiological phenomena, that is to say as being subject to natural and invariable laws, the discovery of which is the special object of its researches'.
- After Saint-Simon and Comte, the Belgian statistician Adolphe Quetelet, proposed that society be modeled using a combination of mathematical probability and social statistics. In fact, Quetelet outlined the project of a social physics characterized by measured variables that follow a normal distribution, and collected data about many such variables (Quetelet 1835).

Perhaps less well known is the other side of the coin that is regarding physics as Nature's economics. For example, economic notions of *value* influenced physicists' attempts to gather empirical evidence for the conservation of energy and formulating the First Law of Thermodynamics. In fact, the work of James Prescott Joule during the 1840s to establish that work and heat are mutually interchangeable and that in every case, a given amount of work would generate the same amount of heat —the so-called *mechanical equivalent of heat* (Joule 1845)—was greatly inspired by the practical experience he collected working as a brewer. Joule, who demonstrated a passion for science since he was a teenager, was the son of a renowned brewer in Manchester. In 1837 he became the manager of the family brewery; however, his passion for science and experimentation continued in parallel with his job. Joule's brewing know-how as well as access to some specialized beer-making equipment would ultimately lead him to discover that heat is a form of energy.

As the historian and philosopher of economic thought Philip Mirowski (1989) points out, three different elements came together in Joule. First, by keeping the books of the brewery Joule was well trained in using detailed accounting systems, including records of timing and of magnitudes that were indispensable to performing final balances. Thus the idea that inputs and outputs must match is likely to have suggested to Joule a metaphorical connection between energy and money and, in turn, help in developing the intuition of the conservation of energy. Second, the control of temperature was absolutely critical in the fermentation process, and so Joule had access to the most sensitive thermometers of the day and he was familiar with control of temperature. Third, the arrival of steam power had virtually mechanized the entire brewing process by 1800, the first industry in which labor was displaced on such a scale. In fact, the Joule's famous paddle-wheel experiment is just a minor variation on the mashing machine that was standard equipment in large breweries in Britain after the turn of the century (Mirowski 1989). In 1845 Joule using this device specified a numerical value for the amount of mechanical work required to produce a unit of heat. He measured the amount of mechanical work needed to raise the temperature of a pound of water by one degree Fahrenheit, and found a consistent value of 819 foot pound force per British thermal unit (4.404 J cal^{-1}). In 1850, Joule published a refined measurement of 772.692 foot-pounds force per British thermal unit (4.150 J cal^{-1}), closer to modern estimates. Previously Joule had shown experimentally that heat could be generated by an electric current and found a value consistent with the above estimates (4.187 J cal^{-1}). By transitivity this convinced him that there was a fundamental connection between quantities of electricity and the mechanical lifting of weights since both exhibit a stable relationship with heat, leading him to postulate the above-mentioned mechanical equivalent of heat.

4.2 Neoclassical economics

One of the pillars of mainstream economics is *neoclassical* economics. Neoclassical economics emerged from the economics as physics approach chronicled in the previous subsection. In neoclassical economics, supply and demand are the driving forces behind the production, pricing, and consumption of goods and services (Encyclopedia Britannica 1998). This conceptual framework was developed in the late 19th century mainly by William Stanley Jevons (1871), Carl Menger (1871), and Léon Walras (1874). Actually, the main founders of neoclassical economics had a background in natural sciences or mathematics. For example, Jevons (1835–92) started his education in natural sciences and Walras (1834–1910) was trained as an engineer. Other important figures in the development of neoclassical theory, like Francis Y Edgeworth (1845–1926) and Alfred Marshall (1842–1924), originally were mathematicians. In fact, the Prussian economist Hermann Heinrich Gossen (1810–58) who is often regarded as the first to elaborate a general theory of marginal utility, although trained in business administration, was very much attracted by mathematics. He laid the mathematical foundation for various insurance forms. Without his mathematical abilities he certainly would not have laid the foundations of marginal utility around 1850, i.e., more than 20 years before Jevons's contribution. Unfortunately, he was not recognized while he was living, since there were few mathematical contributions to economics at his time. Jevons got to know Gossens' ideas only after having finished his work in 1878.

Neoclassical theory emerged to compete with the earlier theories of classical economics which assumed that the most important factor in a product's price is its cost of production. According to neoclassical economics, utility to consumers, not the cost of production, is the most important factor in determining the value of a product or service.

Neoclassical economics rests on three assumptions (Weintraub 2007):

1. People have rational preferences between outcomes that can be identified and associated with values; this is the tenet of *rational choice theory*.
2. People act independently on the basis of full and relevant information.
3. Individuals maximize utility and firms maximize profits, which is the basis of the *marginal theory*.

The first two assumptions can be combined in the so-called *homo economicus* model which portrays humans as ideal decision-makers with complete rationality, perfect access to information, and consistent, self-interested goals.

Let us briefly review this *homo economicus* and the marginal theory.

4.2.1 The *homo economicus* of rational choice theory

Rational choice theory assumes that individuals have preferences among available alternatives and that they perform a cost-benefit analysis to decide whether an option is right for them. For most of the past two centuries, economic thinking has been dominated by the theoretical abstraction of *homo economicus* used to describe in economics a rational human being.

The origins of the *homo economicus* lie in an essay about the political economy by the English civil servant, philosopher, and political economist John Stuart Mill in 1836. Mill's theory was an extension of other ideas proposed by economists, such as Adam Smith and David Ricardo, who also saw humans as primarily self-interested economic agents. Indeed, Smith in his 1776 book *An Inquiry into the Nature and Causes of the Wealth of Nations* characterized humans as motivated by economic self-interest and the maximization of pleasure.

In addition, the *homo economicus* has perfect market information, i.e., all consumers and producers have perfect and instantaneous knowledge of all market prices, their own utility, and own cost functions. This assumption of a perfect market is crucial for *general equilibrium theory*, as we will see in the next subsection.

4.2.2 The marginalist revolution and the general equilibrium theory

The change in economic theory from classical to neoclassical economics, i.e., from the labor theory of value to equate value with *marginal utility*, has been called the *marginal revolution*. The main idea is that the values that people place on economic goods and the prices they set for them are based on the specific uses that people have for each individual unit of a good. People will put the first unit of a good they are able to obtain to their most highly valued use, and use subsequent marginal units for less and less valued ends. This is known as the concept of *diminishing marginal utility*. Marginal utility is derived as the change in utility as an additional unit is consumed. Each person would continue to consume a good or service until the marginal utility would be equal to the price (Hirshleifer and Glazer 1992).

More specifically, marginal theory states that the value of a good or service is determined through a hypothetical maximization of utility by income-constrained individuals and of profits by firms facing production costs and employing available information and factors of production.

It turns out that this is the mathematics of constrained maximization within a conservative field. In short, the marginalists adopted the mathematical formalism of mid-19th-century physics; utility became the analogue of potential energy and the budget constraint became the analogue of kinetic energy (Mirowski 1989).

Walras was able to articulate the utility maximization of the consumer to derive the *laws of supply and demand* which explain how market economies allocate resources and determine the prices of goods and services that we observe in everyday transactions. The law of demand states that **quantity purchased decreases with price**, and vice versa. This occurs because of diminishing marginal utility. That is, consumers use the first units of an economic good they purchase to serve their most urgent needs first, and use each additional unit of the good to serve successively lower-valued ends (Hayes 2021). This can be formalized as the **Law of Diminishing Marginal Utility**, which states that, all else equal, as consumption increases, the marginal utility derived from each additional unit declines. In turn, the law of supply says that **the quantity of goods or services that suppliers offer increases with price**, and vice versa. This is because firms will produce additional output as long as the *marginal cost*—i.e., the cost of producing an extra unit—is less than the market price they receive. That is, as the price of an item goes up, suppliers will attempt to maximize their profits by increasing the quantity offered for sale.

A popular graphical representation of the laws of supply and demand is provided by plotting the price, P, of a product or service on the vertical axis and its quantity, Q, on the horizontal axis (figure 4.1). Note that this implies an exchange of the axis with respect to the canonical representation in which the independent variable appears on the horizontal axis and the dependent one on the vertical axis. Here price, which is the independent variable, appears on the vertical axis and quantity, which is the dependent variable, appears on the horizontal axis. The supply curve is the relationship between the price of a good and the quantity supplied by producers; it parallels the marginal cost curve and thus, as shown in figure 4.1, it moves upward from left to right. The demand curve represents the amount of a certain good that buyers are willing and able to purchase at various

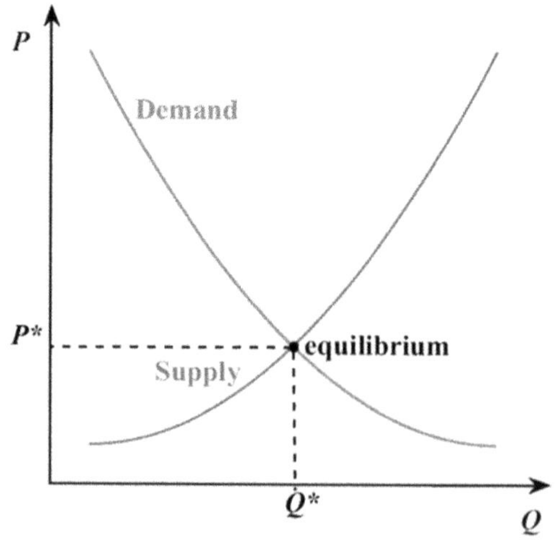

Figure 4.1. Supply and demand curves.

prices, assuming all other determinants of demand are held constant, such as income, tastes and preferences, and the prices of substitute and complementary goods. The demand curve parallels marginal utility and so this curve is always downward-sloping, meaning that as the price decreases, consumers will buy more of the good.

Market equilibrium is reached at a point where the supply and demand curves intersect (black circle in figure 4.1), meaning that the quantity supplied equals demand. The process of moving to a position where the quantity supplied of whatever is traded is equal to the quantity demanded is called *market clearing*. Such market clearing relies upon the assumption that in any given market all buyers and sellers have access to information and that there is no 'friction' impeding price changes, prices always adjust up or down to ensure market clearing.

This idea that supply and demand interact and tend toward a balance in an economy of multiple markets working at once forms the basis of the so-called *general equilibrium theory*. General equilibrium theory attempts to explain the functioning of the macroeconomy as a whole through the balance of competing levels of supply and demand in different markets ultimately creates a price equilibrium (Frankenfield 2021). In fact, general equilibrium theory tried to show how and why all free markets tend toward equilibrium in the long run. However, it is worth remarking that markets didn't necessarily reach equilibrium, only that they tended toward it. As Walras (1874) wrote, 'The market is like a lake agitated by the wind, where the water is incessantly seeking its level without ever reaching it.'

To conclude this section, we note that neoclassical economics has come under considerable question in recent years. One of the most important criticisms is that it relies almost exclusively on the axiomatic-deductive approach while empirical evidence often only plays a minor role. Indeed, models largely function as a substitute for empirical evidence. Thus, in a way it seems that mainstream economics diverged from physics for which information derived from sensory experience, as interpreted through reason and logic, forms the exclusive source of all genuine knowledge. In fact, it has been argued that neoclassical economists have blindly copied the physical theories in the 1870s (Mirowski 1989). In the next section first we will review the main criticisms to neoclassical economics and next we will introduce a different **economics as biology** approach.

4.3 Economics as biology, or evolutionary economics

4.3.1 The irrationality of rational decision theory

History and various economic crises over the years have proved that the theory of a homo economicus is a flawed one. For example, the relatively new branch of *behavioral economics* challenges the traditional view of the homo economicus by providing several evidence of how psychology affects economic decisions. In fact, according to behavioral economists, humans are anything but rational. Behavioral economics attempts to explain, from a psychological perspective, why individual actors sometimes make irrational decisions, and why and how their behavior does not always follow the predictions of economic models (Ganti 2021).

The most important trait of homo economicus is that they are always able to make decisions that allow them to pursue the goal of maximizing profit in the most efficient way. That is, if they are a consumer, the primary goal is to maximize utility; if they are a producer, their primary goal is profit. However, not only are individuals not always self-interested, but they also are not always concerned with maximizing benefits and minimizing costs. Most decision-making occurs with insufficient knowledge and processing capability, and we sometimes lack the self-control to engage in self-interested behavior. In addition, our preferences change, often in response to the context in which a decision is being made (Chen 2021). In other words, the theoretical abstraction of the homo economicus is incompatible with the tenets of behavioral economics.

An important step toward debunking the homo economicus paradigm was provided by the 1978 Nobel Prize in Economics, Herbert Simon. His research was noted for its interdisciplinary nature and spanned across several fields: from cognitive and computer science to public administration, management, and political science. Simon was also among the pioneers of several modern-day scientific domains such as artificial intelligence, information processing, decision-making, problem-solving, organization theory, and complex systems. He rejected the assumption of perfect rationality in mainstream economics and proposed the theory of *bounded rationality* instead. Bounded rationality is a central theme in behavioral economics that sates that people are not always able to obtain all the information they would need to make the best possible decision. On the contrary, knowledge of all alternatives, or all consequences that follow from each alternative, is realistically impossible for most decisions that humans make.

Simon influenced the work of the most conspicuous exponents of behavioral economics, like Daniel Kahneman (Nobel Prize in Economics 2002), Amos Tsversky and Richard Thaler (Nobel Prize in Economics 2017). Indeed, it is argued that Daniel Kahneman and Amos Tversky founded the field of behavioral economics with their 1979 paper, 'Prospect theory: an analysis of decision under risk.' Kahneman and Tversky researched human risk aversion, finding that people's attitudes regarding risks associated with gains are different from those concerning losses. This finding contradicts the prevailing paradigm at this time: homo economicus, who always acts rationally, should not be affected by risk aversion. For example, Kahneman and Tversky found that if given a choice between definitely getting \$1000 or having a 50% chance of getting \$2500, people are more likely to accept the \$1000. The calculation of homo economicus would have been that the average profit is $\frac{1}{2} \times \$2500 - \frac{1}{2} \times \$1000 = +\$750$, and thus accept the bet. In a similar vein, Thaler's idea of mental accounting shows how people place greater value on some dollars than others, even though all dollars have the same value. On a \$20 purchase they might drive to another store to save \$10 but they would not drive to another store to save \$10 on a \$1000 purchase.

Hence, incompatibility between actual behavior of human beings and the behavior of homo economicus has shown the futility of trying to represent real-world systems with such a model flagrantly at odds with reality. Actually, there are alternative models of human decision-making that have been proposed over the years. For instance:

- *Homo reciprocans*, who rewards positive actions and punishes negative actions (Dohmen *et al* 2009, Gintis and Orr 2001).
- *Homo politicus*, that always acts in a way that is consistent with what is best for society (Nyborg 2000).
- *Homo sociologicus*, who is not always perfectly rational because they are affected by society and influenced by societal forces (Hirsch *et al* 1990).

4.3.2 Evolutionary economics

Evolutionary economics proposes that economic behavior is determined both by individuals and society as a whole through economic processes that evolve similarly as by natural selection. The term was first coined by the American economist and sociologist Thorstein Veblen in 1898. Evolutionary economics rejects two of the main assumptions of neoclassical economics:

(a) First, the homo economicus model, arguing that psychological factors are key drivers of the economy;
(b) Second, the notion that economy always tends toward a state of equilibrium; instead evolutionary economists believe the economy is dynamic, constantly changing, and chaotic.

Veblen used an example of social hierarchy and status to illustrate points (a) and (b). He noticed that demand for some goods tends to increase when the price is higher—otherwise known as *conspicuous consumption*, regarded as a means to show one's social status. Conspicuous consumption can be applied to luxury goods that are easily recognizable as high-end, expensive items, like clothing, cars, etc. Veblen (1898) argued that the economic life history of individuals is a cumulative process in which individuals are constantly engaged in the dynamic activity of adapting means to ends within an ever changing environment. In this process, both the individual and their environment are constantly mutually affected. Veblen's concept of the individual has two aspects: instincts, and habitual elements of activity that lead to the creation of institutions. That is, at the heart of human nature lie the instincts, but 'on the other hand the habitual elements of human life change unremittingly and cumulatively, resulting in a continued proliferous growth of institutions' (Veblen 1964). This is very different from the *homo economicus* neoclassical paradigm; man is not a 'lightning calculator of pleasures and pains', but is more akin to a 'coherent structure of propensities and habits which seeks realization and expression in an unfolding activity'. Desire here arises from temperament, which itself is a product of habits and instincts. The environment and the individual could be said to both be emergent properties that affect and are affected by one and another (Howell 2014). In summary, according to Veblen (1961):

'Evolutionary economics must be the theory of a process of cultural growth as determined by economic interest, a theory of cumulative sequence of economic institutions stated in terms of the process itself'.

Expanding on Veblen's early observations Austrian economist Joseph Schumpeter also played an important role in the development of evolutionary economics. Schumpeter's individual is also not based on rational decision-making; rather he believes that the average person is a creature operating under a system in which past experience forms the heart of present action. Individuals will behave then according to the adaptive behavior they have acquired through time and a trial and error process. Furthermore, the Schumpeterian individual is often completely irrational. This is the case of his most well-known conception, the *entrepreneur* (Wunder 2007). Schumpeter paints this entrepreneur with: 'Then there is the will to conquer: the impulse to fight, to prove oneself superior to others, to succeed for the sake, not of the fruits of success, but of success itself. From this aspect, economic action becomes akin to sport—there are financial races, or rather boxing-matches. The financial result is a secondary consideration, or, at all events, mainly valued as an index of success and as a symptom of victory, the displaying of which very often is more 'important as a motive of large expenditure than the wish for the consumers' goods themselves'.

That is, the entrepreneur's behavior is rational in the sense of successfully exploiting the objective possibilities of innovation, yet irrational in that he is ridden by a demon who never lets him be satisfied by results and makes him prone to take high degrees of risk. According to Schumpeter (1911), the entrepreneur is a quite special actor willing to break through traditional structures and to challenge the accepted way of doing things. Entrepreneurs are innovators: people who come up with ideas and embody those ideas in high-growth companies. They continually introduce new goods, services, or new means of production. Entrepreneurs, in his view, could launch a new type or version of a product, like the variety of apps in today's world. Or they can introduce new methods of production—like the *industrial internet of things* (i.e., the combination of interconnected sensors, instruments, and other devices networked together with computers' industrial applications), AI, robotics. In this way Schumpeter's entrepreneur is an agent of change who continually destroys the macroeconomic equilibrium assumed in neoclassical economics by introducing disrupting innovations. Therefore, entrepreneurs are the source of his great *creative destruction.* According to Schumpeter creative destruction describes:

'...the process of industrial mutation that continuously revolutionizes the economic structure from within, incessantly destroying the old one, incessantly creating a new one'.

In other words, the deliberate dismantling of established technologies in order to make way for improved methods of production. Disruptive technologies such as the railroads and Henry Ford's assembly line or, in our own time, the internet are all examples of creative destruction which lead a relentless drive of capitalism toward progress.

Later on, in the 1950s Milton Friedman proposed that the function of markets rests on a 'survival of the fittest' principle. That is, *natural selection* in biology is the mechanism proposed by Darwin to explain evolution: the environment selects those individuals (or what are known as phenotypes) best fitted to survive, while individual less fitted to survive fail to reproduce. Natural selection therefore, is concerned with the differential rates of expansion ('fitness') of the competing, interacting members of the population. In a completely analogous way firms compete for money and unsuccessful rivals who fail to capture an appropriate market share, go bankrupt and have to exit (Friedman 1953).

Evolutionary economics as a field received a boost in 1982 with the publication of *An Evolutionary Theory of Economic Change* by Richard Nelson and Sidney G Winter. These authors conceive firms as a collection of heterogeneous organizations guided by *routines*, the evolutionary economic equivalent of genes. Nelson and Winter reasoned that if the change occurs constantly in the economy, then some kind of evolutionary process must be in action, and there has been a proposal that this process is Darwinian in nature. In fact, they traced a parallelism between phenotypes and companies in a market. Nelson and Winter proposed that over time, the economic analogue of natural selection operates as the market determines which firms are profitable and which are unprofitable, and tends to winnow out the latter (Nelson and Winter 1982). This involves a notion of **fitness of a firm**, reflecting its relative competitiveness, compared to other companies that are competing for the same resource (money, customers) which determines its probability of growth and survival. According to Dosi and Nelson (1994) that fitness is determined by a combination of multiple firm properties like its cash-flow, or its profits, or the expectations that investors hold about future profits, etc.

In the next three sections we will review the basic math used to model Darwinian evolution by natural selection in biology. Section 4.4 is devoted to selection dynamics. Section 4.5 is an interlude to show that: (i) frequency dependent selection equations are actually equivalent to the generalized Lotka–Volterra ecological equations; and (ii) they can be regarded as master equations used in statistical physics. Section 4.6 deals with how the selection equations can be extended to take into account mutations.

4.4 Selection dynamics

4.4.1 The simplest selection model

According to equation (3.3) an isolated biological population will grow exponentially with time as

$$n(t) = n_0 e^{rt}, \tag{4.1}$$

where $n_0 = n(0)$ is the initial population and r denotes its net constant per-capita growth rate. Selection operates whenever different types of individuals growth at different rates. The simplest case is when we have two types. Let us call them 1 and 2. Type 1 individuals growth at rate r_1. Type 2 individuals growth at rate r_2. The rate of growth is interpreted as fitness. Therefore the fitness of 1 is r_1 and the fitness of 2 is r_2. It is simpler to work with relative populations or population fractions (also called population frequencies) instead of populations. This is particularly the case in situations in which the total population size is held constant. Let $x_1(t)$ and $x_2(t)$ denote, respectively, the relative abundance of type 1 and type 2 at time t. Since there are only type 1 and type 2 individuals in the population, we have $x_1(t) + x_2(t) = 1$. We thus have the system of two differential equations:

$$\frac{dx_1}{dt} = x_1(r_1 - \phi)$$
$$\frac{dx_2}{dt} = x_2(r_2 - \phi), \tag{4.2}$$

where the term ϕ ensures that $x_1(t) + x_2(t) = 1$. Thus, summing the two above equations, we get that $0 = r_1 x_1 + r_2 x_2 - (x_1 + x_2)\phi$, and thus

$$\phi = r_1x_1 + r_2x_2. \tag{4.3}$$

That is, ϕ is the weighted average fitness of the population.

Indeed, the two differential equations (4.2) can be transformed into a single differential equation if we denote by $x(t)$ the frequency of type 1 and by $1- x(t)$ the frequency of type 2 both at time t. Thus we get:

$$\frac{dx}{dt} = (r_1 - r_2)x(1 - x). \tag{4.4}$$

Equation (4.4) is the logistic equation (3.5) with growth rate $r_1 - r_2$ and carrying capacity = 1. Hence, it is immediate to see that if rate $r_1 \neq r_2$ equation (4.4) has two equilibria, one for $x = 0$ and the other for $x = 1$, since at these two points, we have $dx/dt = 0$. If $x = 1$ then the system consists only of individuals of type 1, while if $x = 0$ the system consists only of individuals of type 2.

The stability of these two equilibria depends on the sign of $r_1 - r_2$. If $r_1>r_2$, then $dx/dt > 0$ for all values of x that are strictly greater than 0 and strictly smaller than 1. This means that for any mixed system (consisting of some individuals of type 1 and some individuals of type 2) the fraction of subpopulation 1 will increase until it converges to 1. Or, equivalently, the fraction of subpopulation 2 will decrease until it converges to 0. If $r_1<r_2$ then 1 and 2 are exchanged and we end with a population of entirely individuals of type 2 (i.e., $x = 0$ which is equivalent to $1 - x = 1$).

That is, we arrived at *survival of the fitter*.

This model can be extended to describe selection among N different types of individuals. Let us denote by $x_i(t)$ the frequency of type i at time t, in such a way that the structure of the population is thus given by the vector $\mathbf{x}(t) = [x_1,x_2,\ldots,x_N]$, whose elements verify that:

$$\sum_{i=1}^{N} x_i = 1. \tag{4.5}$$

From a geometrical point of view a set of positive real numbers verifying equation (4.5) defines a simplex S_{N-1}. For example, as shown in figure 4.2, a 0-simplex is a point, a 1-simplex is a line segment, a 2-simplex is a triangle, a 3-simplex is a tetrahedron, and so on and so forth. The interior of the simplex is the set of points $\mathbf{x}$ with the property that $x_i>0$ for all $i = 1,\ldots,N$. The face of the simplex is the set of

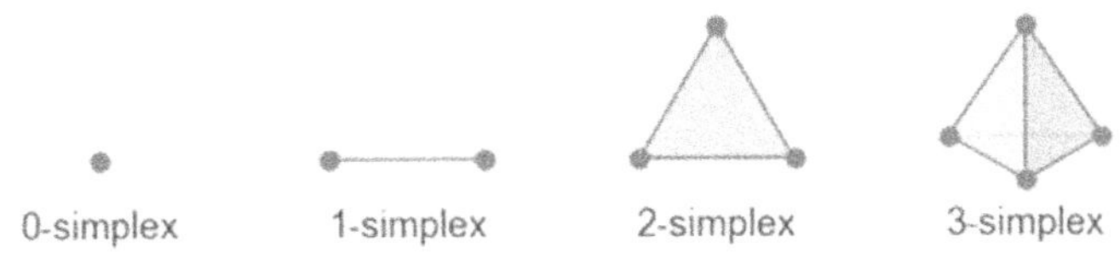

Figure 4.2. The first four simplexes.

points **x** with the property that $x_i = 0$ for at least one i. The vertices of the simplex are the corner points where exactly one type is present, $x_i = 1$, while all other types are extinct, i.e., $x_j = 0$ for all $j \neq i$. The simplex S_1 is given by the *closed interval* [0, 1] (i.e., all real numbers which are greater than or equal to 0 and less than or equal to 1). In contrast, (0, 1) is the *open interval*; it contains all numbers that are strictly greater than 0 and strictly less than 1. The open interval (0, 1) is the interior of the closed interval [0, 1] and, therefore, is also the *interior of the simplex* S_1.

Selection dynamics can be written as:

$$\frac{dx_i}{dt} = x_i(r_i - \phi) \quad i = 1,2, \ldots, N \tag{4.6}$$

with

$$\phi = \sum_{i=1}^{N} x_i r_i, \tag{4.7}$$

i.e., ϕ is, as before, the weighted average fitness of the population. Notice that then equation (4.6) implies that **the frequency of type *i* increases if its fitness exceeds the average fitness of the population**. Otherwise it will decline.

The system of N equation (4.6) contains a single globally stable equilibrium. Starting from any initial condition in the interior of the simplex, the population will converge to a corner point where all but one type have become extinct. This corner corresponds to the type k having the largest fitness, i.e., $r_k > r_i$ for all $i \neq k$. This property is called ***competitive exclusion***: the fittest type will outcompete all others.

However, both in experiments as well as in natural communities coexistence of many different subpopulations is not so rare. How can this be? Notice that here we are considering the simplest linear model; more complex non-linear equations than equation (2.16) can overcome the survival of the fittest outcome (see for instance Nowak 2006). Another interesting possibility to allow for coexistence of different types is associated with the concept of ***frequency dependent selection***. In the next subsection we will show that even in the case of linear models, if the fitnesses of one type depends on the relative abundance of the other types, stable coexistence equilibrium is possible.

4.4.2 Frequency dependent selection: the replicator dynamics equation

The idea of frequency dependent selection is that the fitness of a type i depends on the structure of the population, that is, each species has a *fitness function*:

$$r_i = r_i(\mathbf{x}). \tag{4.8}$$

A popular equation implementing such frequency dependent selection is the *replicator dynamics* (RD) equation introduced in the context of evolutionary game theory (Taylor and Jonker 1978, Schuster and Sigmund 1983). Indeed, several evolutionary models in distinct biological fields—population genetics, population ecology, early biochemical evolution and sociobiology—lead independently to the same class of replicator dynamics (Schuster and Sigmund 1983).

The replicator equation incorporates the dependence on the distribution of N population types through an $N \times N$ *payoff matrix* $\mathbf{P} = [P_{ij}]$. That is, P_{ij} is the payoff for type i when interacting with type j. So the expected payoff of type i is given by

$$\overline{p}_i(\mathbf{x}) = (\mathbf{P}\mathbf{x})_i = \sum_{j=1}^{N} P_{ij}x_j. \tag{4.9}$$

Equating fitness with expected payoff[2], $r_i(\mathbf{x}) = \overline{p}_i(\mathbf{x})$, we substitute equation (4.9) into equation (4.6) and obtain the **replicator dynamics equation** (RDE):

$$\frac{\mathrm{d}x_i}{\mathrm{d}t} = x_i\left(\sum_{j=1}^{N} P_{ij}x_j - \phi\right) i = 1,2,\ldots,N. \tag{4.10}$$

The average fitness ϕ is given by

$$\phi = \sum_{i=1}^{N} x_i r_i(\mathbf{x}) = \sum_{i=1}^{N}\sum_{j=1}^{N} P_{ij}x_i x_j. \tag{4.11}$$

Therefore, the fitness values in the replicator equation (4.10) are linear functions of the frequencies. In other words, the difference between equation (4.10) and equation (4.6) is that of frequency-dependence as opposed to constant selection.

The corners (vertices) of the simplex are fixed points of the replicator dynamics. But, a main novelty of equation (4.10) with respect to equation (4.6) is that now an equilibrium $\mathbf{x}^*$ in the interior of the S_N simplex i.e., $x_i > 0$ for all $i = 1, \ldots, N$ is possible. In fact, depending on the payoff matrix, $\mathbf{P}$, there can be fixed points in the interior and in every face of the simplex S_{N-1} (for a demonstration we refer the reader to Nowak 2006).

Remark: Since the RDE is a deterministic equation each face of the simplex—a subset of the simplex where at least one strategy has zero frequency—is invariant. This is because the RDE is not able to innovate new types through mutations; a type which was not there at the initial time will not appear.

It is instructive to analyze a little closer the RDE for just two types. In this case, as we did before, we can denote the frequency of type 1 as x and the frequency of type 2 as $1 - x$. Thus we can write the average fitness ϕ as:

[2] Incidentally, notice that this implies that the dimensions of the elements of the payoff matrix $\mathbf{P}$ must be time^{-1}.

$$\phi = x r_1(x) + (1 - x) r_2(x). \qquad (4.12)$$

It can be easily verified that equation (4.10) reduces to:

$$\frac{dx}{dt} = x(1 - x)[r_1(x) - r_2(x)]. \qquad (4.13)$$

Therefore, in addition to the two equilibria we found before for the case of both types with fixed fitness, namely $x = 0$ and the other for $x = 1$, there is an additional interior equilibrium x^* that verifies that

$$r_1(x^*) = r_2(x^*). \qquad (4.14)$$

For this interior equilibrium x^* to be stable the polynomial of degree 3 of the right hand side of equation (4.13) must look like figure 4.3. The arrows in this figure to the right or left to each side of each equilibrium point in the direction of the frequency change specified by the sign of the derivative. Notice that arrows point away from 0 or 1 on either but they both point toward x^* from either side. This means that an interior equilibrium, x^* is stable if the derivatives of the functions $r_1(x)$ and $r_2(x)$ satisfy $r_1'(x^*) < r_2'(x^*)$ in such a way that the curve representing the polynomial decreases when around x^*.

In box 4.1 we analyze a timely example showing how the above selection equations can describe the evolutionary dynamics of the ongoing COVID-19.

In summary, the simplest choice for the fitness function r_i(x) is a constant, this is the case we studied in the previous subsection. It leads unavoidably to survival of the fittest. The next simplest choice is a linear dependence of r_i with the frequencies, which corresponds to the RDE. In this case a coexistence equilibrium (a fixed point in the interior of the simplex) is possible. In the case of the rise of Alpha variant discussed in box 4.1 we found examples of both cases.

4.5 Linking selection dynamics with ecology and physics

During ecology's early years, evolutionary thinking was prevalent. However, as ecology focused more and more on abiotic and biotic causes of diversity and species abundance, evolutionary thinking became less prominent (Collins 1986, National Research Council 2005). Much of ecology now operates under the premise that ecological and evolutionary processes act on different timescales. That is,

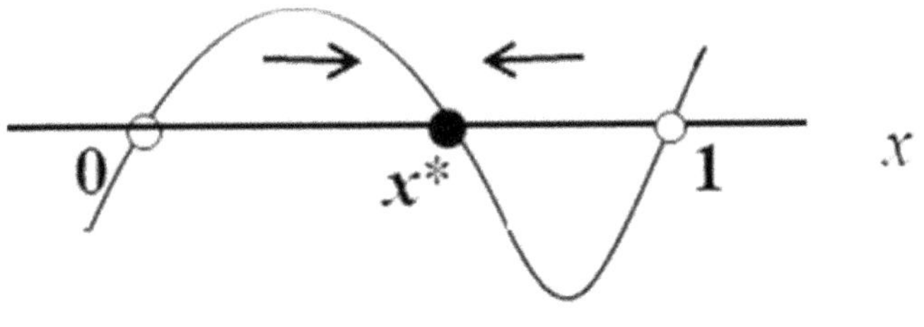

Figure 4.3. Stability of the interior equilibrium for two types. Non-filled circles correspond to unstable equilibria; the filled circle corresponds to a stable interior equilibrium.

Box 4.1. Evolution of the Alpha COVID-19 variant by natural selection in 2020–21 (from Fort 2021) (cont.)

In September 2020 a new variant of the SARS-CoV-2, the virus that causes COVID-19, known as lineage B.1.1.7 (aka Alpha or British variant), was detected in the UK and it quickly displaced the other variants of this virus in several countries (Hodcroft 2021).

It turns out that equation (4.4) can be helpful for describing the change in frequency of B.1.1.7. The idea is to use a very simple model that distinguishes the B.1.1.7 *variant of concern* (VC), that we denote by type 1, from all the others which are treated just as one 'mean' lineage aka 'wild type' (type 2). Thus, r_1 is the growth rate of the VC and r_2 is the growth rate for the *wild type*. If we now divide equation (4.4) by r_2 we get the following differential equation for the frequency of the VC, x:

$$\frac{dx}{dt} = (f - 1)x(1 - x), \tag{i}$$

where $f \equiv r_1/r_2$ is the fitness of this VC relative to the wild type. The solution of the logistic equation (i) is (Fort 2020):

$$x(t) = \frac{1}{1 + \left(\frac{1}{x(0)} - 1\right)e^{-(f-1)t}}. \tag{ii}$$

For long times, if $f > 1$, x will converge to an asymptotic value $x^* = 1$.

Let us compare equation (ii) against empirical data for different European countries. To do this we vary the parameter f in steps of 0.1 and chose the one that produces minimum mean absolute error (MAE). As shown in figure 4.4 this formula can fit the observed change in frequency x of B.1.1.7 until it became 1 or close to 1 for several countries, regions of countries with a single parameter, the relative fitness f of this variant, which is almost universal.

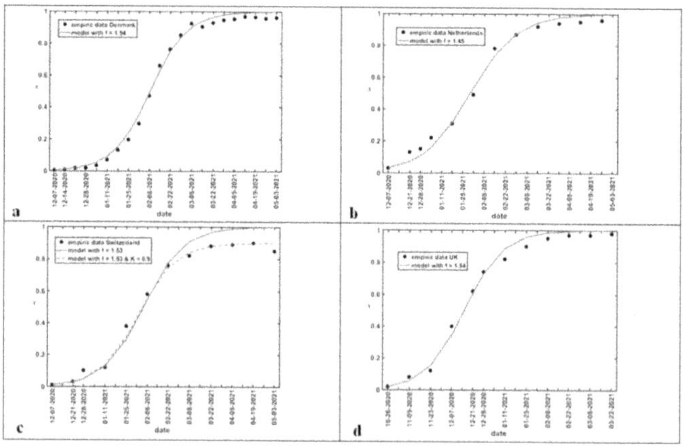

Figure 4.4. Change in the frequency of B.1.1.7 in four European countries (October 2020–May 2021); empirical data (black circles) and formula (ii) (solid line). (a) Denmark. (b) Netherlands. (c) Switzerland. (d) UK.

Notice that this formula in general fits very well with the data, yielding a clean sigmoid curve when plotted versus time. This would be expected under controlled laboratory conditions, but not so much for this kind of measurements involving several non-controlled factors and different kind of heterogeneities.

In the case of Switzerland (panel d) the frequency of B.1.1.7 reached a maximum of $x = 0.9$ and then started to decline. So we also include a fit with an asymptotic carrying capacity $k = 0.9$ (dashed line) to show that the logistic fit for the last part of the sequence greatly improves. Such cases with $x^* = k < 1$ imply that there is an additional equilibrium, i.e., instead of (i) we have:

$$(k - x^*)x^*(1 - x^*) = 0 \tag{iii}$$

Notice that equation (iii) can be obtained from equation (4.13) and thus indicates frequency dependent selection. That is, this interior equilibrium can be interpreted as a situation in which the B.1.1.7 reaches a stable coexistence with some other variants. In this particular case in which $k = 0.9$ these other variants represent a 10% of SARS-CoV-2 population.

Interestingly, the best fit for the four countries occurs for similar values of the relative fitness parameter; around 1.5. Additionally, the values of $f = 1.54$ and $f = 1.53$ for, respectively, Denmark and Switzerland are in agreement with a reported 52% higher transmissibility when compared to the wild type in Denmark and a 51% higher transmissibility when compared to the wild type in Switzerland (ISPM 2021). This suggests that it is right to interpret f as the relative fitness to the local wild type. Furthermore, a 50% higher transmissibility is in agreement with previous estimates reported by Davies *et al* (2020). In summary, we can conclude that:

1. The change in frequency of Alpha variant in time is well described by formula (ii) resulting from an elementary evolutionary model with a single parameter, the relative fitness f. This parameter can be interpreted as the relative fitness to the local wild type.
2. The value of this parameter f varies around 1.50 for all time series considered in this study. This would imply that the variant is 50% more transmissible than the local wild type.
3. The period in which this variant displaces all or most of the other variants has been in all the studied cases around 22 weeks. This is consistent with the above-mentioned almost universal relative fitness f.

evolutionary processes are often thought to take hundreds of generations before their effects can be measured, whereas ecological processes often show effects after a few generations. This has led to the intellectual separation of ecology and evolution. For systems that are under strong selection, however, this may not be the case. A classic example is melanism among moths as a response to air pollution (Kettlewell 1955). An increasing number of studies are combining ecological and evolutionary models to meet this challenge of understanding the consequences of ecological and evolutionary forces acting on similar timescales (Antonovics 1992, Thompson 1999, Whitham *et al* 2003, National Research Council 2005). A very recent example of populations under strong selection is the still ongoing (as of August 2021) pandemic of COVID-19. During this pandemic, vaccines were developed and deployed in real time. Meanwhile, SARS-CoV-2 is replicating in tens of millions of people, some of whom have been immunized, all of whom exert selective pressure for the virus to find new, more efficient replication strategies. The virus will continue to mutate every moment of every day, for years, for decades. Although most genetic changes are innocuous, some can make the mutant more adept at infecting cells, for example, or evading antibodies. Such 'fitter' variants can outcompete other strains, so that they become the predominant source of infections (as shown in box 4.1 for the Alpha variant). A succession of more-transmissible variants has emerged over the past year (September 2020–August 2021). The most worrisome so far (August 2021) is the so-called Delta variant. It has spread to more than 100 countries since it was first reported in India in October 2020. Indeed, this concept of *competitive succession* parallels ecological succession, the process by which the structure of a biological community evolves over time. Let us present in section 4.5.1 a formal connection between evolutionary biology and theoretical ecology. Specifically, we prove the equivalence between the RDE in N-variables and the generalized Lotka–Volterra equations in $N-1$ variables. Next, in section 4.5.2, we show that selection equations, like (4.6) or (4.10), are examples of what in statistical physics is called a ***master equation***. This, in turn, provides an additional bridge connecting evolution with physics.

4.5.1 The equivalence of the replicator dynamics with the generalized Lotka–Volterra equations

We want to prove that the RDE in N-variables $x_1, x_2, \ldots, x_N$ is equivalent to the generalized Lotka–Volterra equations in $N-1$ variables $n_1, n_2, \ldots, n_{N-1}$ (which can represent biomasses or biomass densities, etc). To do this let us start from the Lotka–Volterra equations (3.6) for the $N-1$ variables $n_1, n_2, \ldots, n_{N-1}$, which can be thought as the species yields:

$$\frac{\mathrm{d}n_i}{\mathrm{d}t} = n_i\left(r_i + \sum_{j=1}^{N-1} r_i I_{ij} n_j\right) \quad i = 1, \ldots, N-1. \tag{4.15}$$

Let us define $n_N \equiv 1$ (in the appropriate units, i.e., kg or kg m^{-2}, etc). For coherence of presentation with chapter 3 it is convenient to introduce non-dimensional variables y_i defined as $y_i \equiv n_i/n_N$. In terms of these new variables equation (4.14) can be rewritten as:

$$\frac{\mathrm{d}y_i}{\mathrm{d}t} = y_i\left(r_i + \sum_{j=1}^{N-1} r_i I'_{ij} y_j\right) \quad i = 1, \ldots, N-1, \tag{4.15'}$$

where the prime in the interaction matrix is to remember that now it is a non-dimensional quantity (while I_{ij} has dimensions of the inverse of variable n). We can obtain the frequencies x_i as:

$$x_i = \frac{y_i}{\sum_{j=1}^{N} y_j} \quad i = 1, \ldots, N. \tag{4.16}$$

Equation (4.16) can be thought as a transformation from y to x. The inverse transformation, from x to y is given by:

$$y_i = \frac{y_i}{y_N} = \frac{x_i}{x_N} \quad i = 1,\dots,N. \tag{4.17}$$

Using equation (4.17) we can express the time derivative of y_i as

$$\frac{\mathrm{d}y_i}{\mathrm{d}t} = \frac{\mathrm{d}}{\mathrm{d}t}\left(\frac{x_i}{x_N}\right) = \frac{1}{x_N}\frac{\mathrm{d}x_i}{\mathrm{d}t} - \frac{\mathrm{d}x_N}{\mathrm{d}t}\frac{x_i}{{x_N}^2}. \tag{4.18}$$

Thus, substituting equation (4.18) into equation (4.15) we obtain:

$$\frac{\mathrm{d}x_i}{\mathrm{d}t} = x_i\left[\frac{1}{x_N}\frac{\mathrm{d}x_N}{\mathrm{d}t} + r_i + \sum_{j=1}^{N-1} r_i\frac{I'_{ij}}{x_N}x_j\right]. \tag{4.19}$$

Now if we identify the term in the parenthesis of equation (4.19) with the one in the parenthesis of equation (4.10) we get:

$$\frac{1}{x_N}\frac{\mathrm{d}x_N}{\mathrm{d}t} + r_i + \sum_{j=1}^{N-1} r_i\frac{I'_{ij}}{x_N}x_j = \sum_{j=1}^{N} P_{ij}x_j - \phi, \tag{4.20}$$

which using the RDE for the frequency of x_N transforms into:

$$r_i + \sum_{j=1}^{N-1} r_i\frac{I'_{ij}}{x_N}x_j = \sum_{j=1}^{N}(P_{ij} - P_{Nj})x_j. \tag{4.21}$$

Therefore, the relationship between L–V parameters and RDE parameters is:

$$r_i = P_{iN} - P_{NN}, \tag{4.22a}$$

$$r_i I'_{ij} = (P_{ij} - P_{Nj})x_N. \tag{4.22b}$$

We can take $P_{Nj} = 0$ for all j, this implies no restriction in generality (Hofbauer and Sigmund 1998) and, if we simultaneously redefine the intrinsic rate of growth by absorbing the factor $1/x_N$, we finally obtain:

$$P_{ij} = \begin{cases} r_i I'_{ij} \text{ if } i, j = 1,\dots,N-1 \\ r_i \text{ if } j = N \\ 0 \text{ if } i = N \end{cases} . \tag{4.23}$$

And this completes the proof of the equivalence between the RDE in N-variables x_1, $x_2,\dots,x_N$ defining a simplex S_{N-1} and the Lotka–Volterra equations in $N-1$ variables $y_1,y_2,\dots,y_{N-1}$.

This is a very important outcome since whatever result holds for one equation will hold for the other. The equivalence of the Lotka–Volterra and replicator equations establishes a useful conceptual bridge between evolutionary biology and theoretical ecology. We will use this bridge in chapter 7 to approach markets as ecosystems.

4.5.2 The selection equations regarded as master equations

In the previous subsection we arrived to a powerful conceptual result that allows connecting evolution equations with population equations ruling the dynamics of ecosystems. Now we will turn our attention to statistical physics. Indeed, equations (4.6) or (4.10) are examples of what in physics is called a ***master equation***. Given a system modeled as being in a probabilistic combination of states at any given time, a master equation is a differential equation which expresses the time evolution of this probability distribution of the possible states of the system in terms of the microscopic processes involved. As we have seen in chapter 2, the most classical situation where stochastic processes enter in the description of physical systems is statistical mechanics. Statistical mechanics studies systems of large numbers of particles. Therefore, instead of microscopic precise descriptions it deals with probabilistic descriptions of the microstates to compute average values. The master equation connects the change of the probability of being in a given state with the probabilities of transition to and from any other state in the system.

The selection equations (4.6) and (4.10) are indeed a set of first-order differential equations that describe the time evolution of the probability of occupying a specific set of states.

According to this view, the fitness functions $r_i(\mathbf{x})$ describe the observed transition rates to 'state' i. The functional form of $r_i(\mathbf{x})$ depends on the specific types of evolutionary interactions. We have seen the two simplest examples, namely $r_i(\mathbf{x})$ equal to a constant (independent of $\mathbf{x}$) and a linear function of $\mathbf{x}$ (i.e., the RDE).

The master equation is an approach to stochastic systems applied when time is continuous. Thus, it is the continuous time version of the recurrence relations for a ***Markov chain***, a stochastic model describing a sequence of possible events in which the probability of each event depends only on the state attained in the previous event. Indeed, it is easier to develop the master equation starting from the discrete time case. If we denote the state of the system by s, we can write a difference equation for a particular probability at time t, $P(s;t)$, of the form:

$$P(s;\, t) = P(s;\, t-1)\left[1 - {\sum_{s'}}' P(s \to s')\right] + {\sum_{s'}}' P(s';\, t-1)P(s' \to s), \quad (4.24)$$

where the prime on the sum indicates that the term $s' = s$ is omitted, and $P(s{\to}s')$ is the transition probability from state s' to state s. Notice that since we are assuming a Markov process $P(s{\to}s')$ is time-independent, reflecting that the transition probability does not depend upon the past history that the system has been before state s.

Equation (4.24) can be rewritten as

$$P(s;\, t) - P(s;\, t-1) = {\sum_{s'}}' [P(s';\, t-1)P(s' \to s) - P(s;\, t-1)P(s \to s')]. \quad (4.25)$$

To write the continuum form we take the time difference between steps as Δt instead of 1, and, dividing equation (4.25) by Δt, we get:

$$\frac{P(s;\, t) - P(s;\, t-\Delta t)}{\Delta t} = {\sum_{s'}}' \left[\frac{P(s' \to s)}{\Delta t} P(s';\, t-\Delta t) - \frac{P(s \to s')}{\Delta t} P(s;\, t-\Delta t)\right]. \quad (4.26)$$

Taking the limit $\Delta t \to 0$, equation (4.26) transforms into:

$$\frac{\mathrm{d}P(s;\, t)}{\mathrm{d}t} = \sum\nolimits'_{s'} [w(s' \to s)P(s';\, t) - w(s \to s')P(s;\, t)], \tag{4.27}$$

where the ratio $P(s{\to}s')/\Delta t$ has been replaced by the rate of transition $w(s{\to}s')$. Equation (4.27) is called the Master equation.

The Master equation thus says that:

The rate of change of the probability of a particular state is the total rate at which probability is being added into that state from all other states, minus the total rate at which probability is leaving the state.

An analogy from physics comes to mind: probability acts like a fluid that is flowing to or from a particular state and is being conserved. Equation (4.27) is very much like the continuity equation of fluid flow, where the density of the fluid at a particular place changes according to how much is flowing to that location or from it.

Up to now we have been considering deterministic differential equations for the change in time of populations of individuals; for example the RDE or the formally equivalent generalized LVE. In fact, the dynamics of real populations is stochastic; the abundance of individuals is given by integers rather than by continuous variables that fluctuate due to random births and deaths, migrations and random variations of the physical environment they inhabit. It turns out that the deterministic description is exact only when there are an infinite number of individuals and for realistic population sizes stochastic effects may be important. Therefore, although usually differential equations are easier to analyze and interpret than stochastic processes, many important effects in real systems only arise in a stochastic context.

In box 4.2 we present an example of the use of a master equation in the context of population dynamics proposed by McKane and Newman (2004).

4.6 Innovation through mutations

4.6.1 Evolution as a two-step mutation–selection process

Evolution by natural selection is a two-step process. The first step involves genetic variation within a population. One of the main causes of variation comes by mutations in the genome (the others are reshuffling of genes through sexual reproduction and migration between populations, aka gene flow). The second step determines which randomly generated variants will persist into the next generation. Whereas the origin of a new genetic variant occurs at random in terms of its effects on the organism, the probability of it being passed on to the next generation is absolutely non-random due to its effects on survival and reproduction of that organism (Mayr 2001, Gregory 2009). In the previous two sections we have focused on this second step. Now we turn our attention to mutations.

Importantly, mutations are known to be random or 'undirected' with respect to any effects that they may have. That is, any given mutation is merely a chance error in genetic replication, and as such, its likelihood of occurrence is not influenced by whether it will turn out to be detrimental, beneficial or neutral (Gregory 2009). In

Box 4.2. An example of a master equation for individuals of a species competing for resources and the associated deterministic equation

Let us consider the competition of individuals of one species that we call A which reproduce asexually, like bacteria, in a culture dish. We denote by N the maximum sustainable population or *carrying capacity*. Hence, we model the interactions of A individuals by subdividing the total area of the dish into N plots of equal area. The plot sizes are chosen so that each one either contains one A individual, or it is empty and labeled by E. The population dynamics can be essentially described by three processes: birth, death, and competition. The first and third processes will involve two individuals:

$$AE \xrightarrow{b} AA \text{ (birth)}$$
$$A \xrightarrow{d} E \text{ (death)}$$
$$AA \xrightarrow{c} AE \text{ (competition)}$$

In other words, we are modelling death as constant, independent of the density of individuals, but the reduction in the numbers of A due to competition and the growth in numbers of A due to births will be density dependent. This means that there will be a tendency for AA to go to AE because of overcrowding, and for AE to go to AA due to the presence of resources to sustain a new individual. To obtain the time evolution of this population we need the rates at which the three processes occur. We will assume that for μ of the time randomly choose two individuals and allow them to interact. For $(1 - \mu)$ of the time choose only one individual randomly. Thus if n is the population of species A, and, by simple combinatory we have that:

$$\text{probability of picking } AE = 2\mu\frac{n}{N}\frac{N-n}{N-1},$$
$$\text{probability of picking } A = (1-\mu)\frac{n}{N},$$
$$\text{probability of picking } AA = \mu\frac{n}{N}\frac{n-1}{N-1},$$

where the factor of 2 in the first term comes from the fact that the choices AE and EA are identical. These results enable us to write down expressions for the transition probability, per unit time step, of the system of individuals going from a state with n A individuals to a state with n' A individuals $P(n \rightarrow n')$. Since only transitions from n to $n \pm 1$ may take place during one time step, the only nonzero $P(n \rightarrow n')$ are

$$P(n-1 \rightarrow n) = \mu b\frac{n}{N}\frac{N-n}{N-1}, \quad P(n \rightarrow n-1) = (1-\mu)d\frac{n}{N} + \mu c\frac{n}{N}\frac{N-n}{N-1}. \tag{i}$$

The process defined by equations (i) is a one-step Markov process and so we can immediately write down a master equation describing how the probability of having n individuals, $P(n, t)$, changes with time. This equation is:

$$\begin{aligned}\frac{\mathrm{d}P(n,t)}{\mathrm{d}t} &= P(n+1 \rightarrow n)P(n+1,t) + P(n-1 \rightarrow n)P(n-1,t) \\ &\quad - P(n \rightarrow n-1)P(n,t) - P(n \rightarrow n+1)P(n,t).\end{aligned} \tag{ii}$$

This set of coupled equations has to be solved subject to an initial condition, typically $P(n, 0) = \delta n,n0$, i.e., a condition stating that there are n_0 individuals at $t = 0$. If we define that $P(0\to -1) = 0$ and $P(n\to n+1) = 0$ we can use the general form (ii), even for $n = 0$ and $n = N$, as a complete description of the time evolution.

To make contact with the deterministic version of the model, we use the *mean-field* approximation, obtained by taking averages and the $N \to \infty$ limit. Thus, by multiplying the master equation (ii) by n and summing over all values of n (by shifting the variable in two of the sums on n by +1 and −1) we get:

$$\frac{\mathrm{d}}{\mathrm{d}t}\left(\sum_{n=0}^{N} nP(n,t)\right) = \sum_{n=0}^{N} P(n+1\to n)P(n,t) - \sum_{n=0}^{N} P(n\to n-1)P(n,t). \quad \text{(iii)}$$

Denoting the average of an arbitrary quantity X over the possible states of the system as $\langle X\rangle = \sum_{n=0}^{N} XP(n,t)$ we can re-write equation (iii) as:

$$\frac{\mathrm{d}\langle n\rangle}{\mathrm{d}t} = \langle P(n+1\to n)\rangle - \langle P(n\to n-1)\rangle. \quad \text{(iv)}$$

Using equation (i) and defining

$$\bar{b} = \frac{\mu b}{N-1}, \quad \bar{d} = \frac{(1-\mu)d}{N-1}, \quad \bar{c} = \frac{\mu c}{N-1}, \quad \text{(v)}$$

equation (iv) becomes:

$$\frac{\mathrm{d}}{\mathrm{d}t}\frac{\langle n\rangle}{N} = 2\bar{b}\left\langle \frac{n}{N}\left(1-\frac{n}{N}\right)\right\rangle - \bar{d}\left\langle\frac{n}{N}\right\rangle - \bar{c}\left\langle\frac{n}{N}\left(\frac{n}{N}-\frac{1}{N}\right)\right\rangle. \quad \text{(vi)}$$

So far no approximation has been made in the derivation of equation (vi). We now take the limit $N\to\infty$. In addition to eliminating the $1/N$ factor in the last term on the right hand side of equation (vi), it allows us to replace $\langle n^2\rangle$ by $\langle n\rangle^2$ (this is because the variance of n, $\mathrm{var}(n) \equiv \langle n^2\rangle - \langle n\rangle^2$ is proportional to $1/N$). So, if we identify $n_A(t) = \langle n\rangle(t)$ this gives:

$$\frac{\mathrm{d}n_A}{\mathrm{d}t} = n_A\left(2\bar{b} - \bar{d} - \frac{2\bar{b}+\bar{c}}{N}n_A\right), \quad \text{(vii)}$$

Which reduces to the logistic equation, equation (3.5) provided we define the growth rate r and the carrying capacity K by:

$$r = 2\bar{b} - \bar{d} \quad \text{and} \quad K = \frac{2\bar{b}-\bar{d}}{2\bar{b}+\bar{c}}N. \quad \text{(viii)}$$

In chapter 3 we arrived at the logistic equation as a phenomenological description of the population growth of a single species with intraspecific competition. Here it is derived as the $N\to\infty$ limit of the proposed stochastic model and thus provides a reasonable description of the system when its size (number of A plus number of E types) of the system is relatively large. Of course, this limit is purely formal. In practice populations are finite. For instance, if N is of the order 10^4, then this approximation is good if we are only interested in accuracies of up to 0.01% (if the next-order corrections are of order $1/N$) or 1% (if the next-order corrections are of order $1/\sqrt{N}$). This approximation obviously cannot describe chance extinctions, which occur when n is small, nor does it predict a mean time to extinction for the A population.

fact, most mutations are neutral and from those mutations that have an impact on survival and reproductive output the majority are deleterious, and thus will be less likely than existing alternatives to be passed on to subsequent generations. However, a small percentage of new mutations will turn out to have beneficial effects in a particular environment and will contribute to an elevated rate of reproduction, i.e., an increase in fitness, by organisms possessing them. Even a very slight advantage can be amplified by selection and thus causing new beneficial mutations to increase over the span of many generations.

4.6.2 Mutation–selection equations: from the Crow–Kimura equation to the replicator–mutator equation

As we have seen in section 4.4 equations (4.6) and (4.10) model the second step of evolution, i.e., selection; the first corresponds to the frequency independent case and the second to the frequency dependent case. Let us now extend these equations to model replication which also include mutations. A common type of selection–mutation equations is one that uses a discrete *state space* aka *sequence space* (Nowak 2006) describing all the possible (geno)types. Thus mutations occur but only between types that are already present. Selection in turn occurs as these varieties reproduce and compete at different rates depending on fitness. This is implemented by introducing a *mutation matrix* $\mathbf{Q} = [Q_{ij}]$, whose elements q_{ij} denote the probability that type i mutates to type j. Since each type i has to produce itself or some other type, implies that the sum of each row must be 1:

$$\sum_{j=1}^{N} Q_{ij} = 1. \tag{4.28}$$

Thus $\mathbf{Q}$ is a *stochastic N×N matrix*. A stochastic matrix is defined by the properties that: (i) all entries are numbers from the interval [0, 1] (so-called probabilities); (ii) there are as many rows as columns; and (iii) the sum of each row is 1, i.e., equation (4.28).

A limitation of these models is that they do not provide direct insight into how a population forms new types that may be more fit or less fit than the original population, i.e., how innovation really occurs (in the genetic parlance, no genes are lost and no new genes are created). However this kind of modelling is important for understanding the selection–mutation process as a population adapts to its environment.

We begin by considering two mutation–selection equations that generalize frequency independent selection of equation (4.6). The first of these equations was proposed by Crow and Kimura (1970), it is given by:

$$\frac{\mathrm{d}x_i}{\mathrm{d}t} = x_i(r_i - \phi) + \sum_{j=1}^{N} x_j Q_{ji} - x_i \quad i = 1,2,\ldots,N \tag{4.29}$$

Two remarks are in order regarding the terms on the left hand side of Crow–Kimura equation (4.28):

- the first term is just equation (4.6);
- the third term, $-x_i$, ensures that $\sum_{i=1}^{N} \mathrm{d}x_i/\mathrm{d}t = 0$ which is necessary by the constraint of equation (4.5), i.e., $\sum_{i=1}^{N} x_i = 1$.

An alternative mutation–selection equation is the ***quasispecies equation***. A quasispecies is a large group or 'cloud' of related genotypes that exist in an environment of high mutation rate but at stationary state (Eigen *et al* 1989), where a large fraction of offspring are expected to contain one or more mutations relative to the parent. This is in contrast to a species, which from an evolutionary perspective is a more-or-less stable single genotype, most of the offspring of which will be genetically accurate copies (Biebricher and Eigen 2006). The quasispecies equation is given by:

$$\frac{\mathrm{d}x_i}{\mathrm{d}t} = \sum_{j=1}^{N} x_j r_j Q_{ji} - \phi x_i, \qquad i = 1,2,\ldots,N \tag{4.30}$$

Exactly as in the selection equation (4.6), each sequence is removed at rate ϕ to ensure the constraint of equation (4.5), i.e., $\sum_{i=1}^{N} x_i = 1$.

Notice that in the limiting case of completely error-free replication, $\mathbf{Q}$ becomes the identity matrix: all diagonal entries are one, all off-diagonal entries are zero, and thus both the Crow–Kimura equation (4.29) and the quasispecies equation (4.30) reduce to the selection equation (4.6).

We conclude this subsection with an equation that combines frequency dependent selection and mutations; the so-called ***replicator–mutator*** equation (RME), which is:

$$\frac{\mathrm{d}x_i}{\mathrm{d}t} = \sum_{j=1}^{N} x_j r_j(\mathbf{x}) Q_{ji} - \phi x_i = \sum_{j=1}^{N}\sum_{k=1}^{N} P_{jk} Q_{jki} x_j x_k - \phi x_i, \qquad i = 1,2,\ldots,N \tag{4.31}$$

This equation is a simultaneous generalization of the replicator equation (4.10) and the quasispecies equation (4.30). The RME is used in the mathematical analysis of language (see Nowak 2006 and references therein) and in evolutionary economics (Safarzynska and van den Bergh 2008).

The fact that $\sum_{i=1}^{N} x_i = 1$ is obeyed by equations (4.29), (4.30) and (4.31) implies that the three are defined on the simplex, S_N (like selection equations).

4.7 Implementing evolution in economics

As we have seen in section 4.3, Nelson and Winter (1982) took the ideas expressed by Veblen and Schumpeter and presented an evolutionary approach to economic growth, technological progress and competition between firms by focusing mostly on the issue of changes in technology and routines. Because there is nothing which guarantees, in general, the optimality of these routines, notional opportunities for the discovery of 'better' ones are always present. Changes in response to learning processes involving imperfect adaption and mistake-ridden discoveries can be interpreted as 'mutations'. This applies equally to the domains of technologies, behaviors and organizational setups (Nelson and Dosi 1994). Indeed, a case study from the soda and chlorine industry in the 19th century (Faber *et al* 2021), shows that long-term optimization does not exist in reality, yet due to the occurrence of novelty, there exists, at most, myopic (i.e., short-sighted) optimization[3].

To improve their profits, firms search for innovative (or imitative) solution, with successful firms growing at the expense of the less successful. The process is fundamentally dynamic, as firms interact and create their relative competitive environment. In this respect Metcalfe (1998) clearly illustrates the concept of *economic fitness*:

'Economic fitness is a measure of rates of expansion and decline of activity and, since it applies to the business unit, it is partly determined by the capabilities and intention of that unit. However, the crucial property of economic fitness is that it is not a property of the business unit alone, but arises from the interaction between rival business units in a given market environment. It is inherently a feature arising from membership of that particular population. It is caused by the interaction between the individual business units and those populations cum environmental relationships;'

This approach serve to connect the perspective of economics as a complex evolving system with the concepts and models of biological evolution by natural selection (sections 4.4 and 4.6). According to Nelson and Winter, as the market evolves, the market shares of the inefficient companies decrease until they are driven out of the market (Beker 2004). Therefore, as a result of selection, the companies with greatest fitness capture market share.

Developments within evolutionary modelling of economics include various mathematical techniques, such as non-linear dynamic analysis (in terms of both difference and differential equations), stochastic processes and evolutionary algorithms (see Safarzynska and van den Bergh 2008 and references therein). Dynamics focused on selection predominantly employs replicator dynamics. For example, in the Silverberg–Dosi–Orsenigo model (1988), who 'wins' and who 'loses' is determined by a selection process captured by a replicator-type dynamics where market shares change according to the relative values of a vector of characteristics, synthetically called 'competitiveness'. Other stochastic and deterministic selection equations have been utilized, such as imitation, best response, mutator and adaptive dynamics, but they are more rarely used in economic applications.

[3] We encourage the reader interested in evolutionary economics to visit and explore the website of the *Mapping the Interplay between Nature and Economy* (MINE) project, created by Malte Faber at the Alfred-Weber-Institute, Heidelberg University http://nature-economy.de/.

4.8 The 'Marshall problem' or a transdisciplinary synthetic perspective of economics

The British economist Alfred Marshall has been acknowledged as a leading architect of neoclassical economics. As we have seen, neoclassical theory built mainly upon a mechanical metaphor drawn from Newtonian mechanics and later 19th century physics. Neoclassical theory is presented today in standard introductory college textbooks as the foundation of economic theory.

Despite his reliance upon the mechanical metaphor, Marshall has the intuition that economics dealt with a more complex reality, requiring a more complex dynamic. He stated in the Preface of his *Principles of Economics* (1959) that 'the Mecca of economics lay in biology rather than economic mechanics'. In addition, behavioral economists made a case demonstrating that individuals are not just utility maximizers. Instead people are often moved by emotions and risk aversion, etc. This does not mean that *all* neoclassical theory is wrong. Far from it. The value of neoclassical economics can be assessed in the collection of its truths. One of such truths is the undeniable major role of supply and demand in price formation. So when a company wants to introduce a fresh product into the market and wants to find the right price for its product the company performs a survey to measure the demand for the product at different prices, generating a demand curve. This curve allows calculating the profits in each case and the company selects the price where it makes the highest profits. Likewise, the **Law of Diminishing Marginal Returns**—related to the concept of diminishing marginal utility—is used every day by firms to decide whether to hire or not new staff. This law says that at a certain point, employing an additional factor of production causes a relatively smaller increase in output. For example, a factory employs workers to manufacture its products, and, at some point, the company operates at an optimal level. With all other production factors constant, adding additional workers beyond this optimal level will result in less efficient operations. Actually, the law of diminishing returns is not only a fundamental principle of economics, but it also plays a starring role in production theory (i.e., the study of the economic process of converting inputs into outputs) (Hayes 2021). Another example is when planning for future electricity needs in a state, 'for example, the Public Utilities Commission develops a (neoclassical) demand forecast, joins it to a (neoclassical) cost analysis of generation facilities of various sizes and types (e.g., an 800 megawatt low-sulfur coal plant), and develops a least-cost system growth plan and a (neoclassical) pricing strategy for implementing that plan' (Weintraub 2007).

In fact, a profound interconnection between mechanics and biology can be observed in Marshall's work (Cassata and Marchionatti 2011). Importantly, Marshall has been also recognized as the first who addressed the problem of integrating physics and biology simultaneously into economics in a way that both contains its orthodox core as well as provides a transdisciplinary perspective on economic complexity. Finding a way to integrate physics and biology into economics has been called the 'Marshall's problem' (Rosser 2010).

Inspired by Marshall's epistemological and methodological legacy on the issue of complexity of economic systems, in chapter 7 we will combine elements of physics and biology to develop a method of market forecasting. To do this we first have to explain in the next chapter how to infer from datasets the strengths of phenomenological interactions through MaxEnt principle.

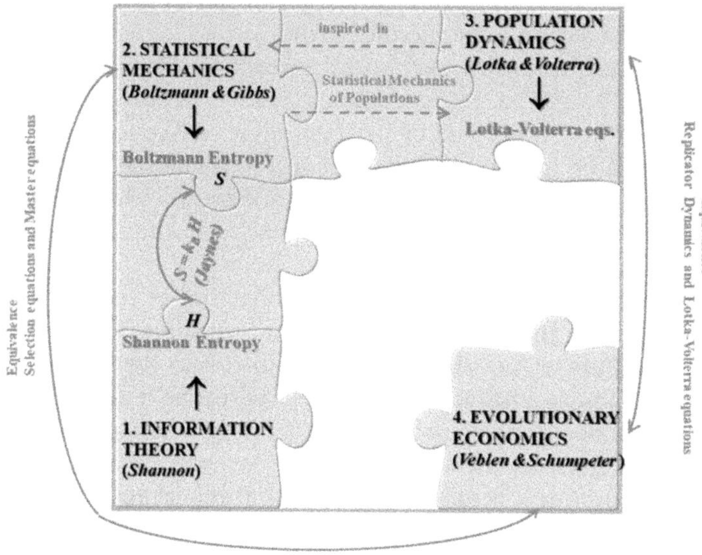

References

Antonovics J 1992 Toward community genetics *Plant Resistance to Herbivores and Pathogens: Ecology, Evolution, and Genetics* ed R S Frite and E L Simms (Chicago, IL: University of Chicago Press) pp 426–49

Beker P F 2004 Are inefficient entrepreneurs driven out of the market? *J. Econ. Theo* **114** 329–44

Biebricher C K and Eigen M 2006 What is a quasispecies *Quasispecies: Concept and Implications for Virology* ed E Domingo (Berlin: Springer)

Cassata F and Marchionatti R 2011 A transdisciplinary perspective on economic complexity. Marshall's problem revisited *J. Econ. Behav. Organ.* **80** 122–36

Chen J 2021 Homo economicus *Investopedia* https://investopedia.com/terms/h/homoeconomicus.asp (accessed 29 July 2021)

Collins J P 1986 Evolutionary ecology and the use of natural selection in ecological theory *J. Hist. Biol* **19** 257–88

Crow J and Kimura M 1970 *An Introduction to Population Genetics Theory* (New York: Harper and Row)

Davies N *et al* 2020 Estimated transmissibility and severity of novel SARS-CoV-2 lineage B.I.I.7 in England *Science* **372** 149

Dohmen T, Falk A, Huffman D and Sunde U 2009 Homo reciprocans: survey evidence on behavioural outcomes *Econ. J.* **119** 592–612

Dosi G and Nelson R R 1994 An introduction to evolutionary theories in economics *J. Evol. Econ.* **4** 153–72

Eigen M, McCaskill J and Schuster P 1989 Molecular quasi-species *J. Phys. Chem.* **92** 6881–91

Encyclopedia Britannica 1998 Alfred Marshall https://britannica.com/biography/Alfred-Marshall

Faber M *et al* 2021 *Mapping the Interplay between Nature and Economy* http://nature-economy.de/ accessed 25 August 2021

Fort H 2021 A very simple model to account for the rapid rise of the alpha variant of SARS-CoV-2 in several countries and the world *Virus Res.* **304** 198531

Fort H 2020 *Ecological Modelling and Ecophysics: Agricultural and Environmental Applications* (Bristol: IOP Publishing) ch 0 pp 0–8

Frankenfield J 2021 General equilibrium theory *Investopedia* https://investopedia.com/terms/g/general-equilibrium-theory.asp accessed 29 July 2021

Friedman M 1953 The methodology of positive economics *Essays in Positive Economics* (Chicago, IL: University of Chicago Press) pp 3–43

Ganti A 2021 Rational choice theory Investopedia https://investopedia.com/terms/r/rational-choice-theory.asp accessed 29 July 2021

Gintis H and Orr H A 2001 Why do we cooperate? *Boston Review* (East Lansing, MI: Michigan State University Press) (Online) http://bostonreview.mit.edu/bostonreview/BR23.1/Gintis.html

Gregory T R 2009 Understanding natural selection: essential concepts and common misconceptions *Evol. Edu. Outreach* **2** 156–75

Hayes A 2021 Law of demand *Investopedia* https://investopedia.com/terms/l/lawofdemand.asp accessed 29 July 2021

Hirsch P, Michaels S and Friedman R 1990 Clean models vs. dirty hands why economics is different from sociology *Structures of Capital: The Social Organization of the Economy* ed S Zukin and P DiMaggio (Cambridge: Cambridge University Press)

Hirshleifer J and Glazer A 1992 *Price Theory and Applications* (Englewood Cliffs, NJ: Prentice-Hall)

Hodcroft E B 2021 CoVariants: SARS-CoV-2 mutations and variants of interest https://covariants.org accessed 21 June 2021

Hofbauer M and Sigmund K 1998 *Evolutionary Games and Population Dynamics* (Cambridge: Cambridge University Press)

Howell T 2014 Selected critiques of the philosophical underpinnings of the neoclassical school *University Honors Theses* Paper 59

Iggers G G 1959 Further remarks about early uses of the term 'Social Science' *J. Hist. Ideas* **20** 433–6

ISPM 2021 Transmission of SARS-CoV-2 variants in SwitzerlandBern Institute of Social and Preventive Medicine (ISPM), University of Bern (retrieved 30 March 2021) https://ispmbern.github.io/covid-19/variants/

Jevons W S 1871 *The Theory of Political Economy* (London: Macmillan)

Joule J P 1845 On the mechanical equivalent of heat *Notices and Abstracts of Communications to the British Association for the Advancement of Science*

Joule J P 1850 On the mechanical equivalent of heat *Phil. Trans. R. Soc.* **140** 61–82

Kahneman D and Tversky A 1979 Prospect theory: an analysis of decision under risk *Econometrica* **47** 263–91

Kettlewell H B D 1955 Selection experiments on industrial melanism in the Lepidoptera *Heredity* **9** 323–42

Mantegna R N and Stanley H E 2000 *An Introduction to Econophysics* (Cambridge: Cambridge University Press)

Marshall A 1959 [1890] *Principles of Economics* (New York: Macmillan)

Mayr E 2001 *What Evolution Is* (New York: Basic Books)

McKane A J and Newman T J 2004 Stochastic models in population biology and their deterministic analogs *Phys. Rev.* E **70** 041902

Menger C 1871 *Principles of Economics (German: Grundsätze der Volkswirtschaftslehre)* (Wien: Braumüller Pub.)

Metcalfe J S 1998 *Evolutionary Economics and Creative Destruction* (London: Routledge)

Mill J S 1836 *The Collected Works of John Stuart Mill, Volume IV—Essays on Economics and Society Part* I https://oll.libertyfund.org/title/mill-the-collected-works-of-john-stuart-mill-volume-iv-essays-on-economics-and-society-part-i accessed 12 May 2021

Mirowski P 1989 *More Heat than Light: Economics as Social Physics, Physics as Nature's Economics* (Cambridge: Cambridge University Press)

National Research Council 2005 *Mathematics and 21st Century Biology* (Washington, DC: The National Academies Press)

Nelson R R and Dosi G 1994 An introduction to evolutionary theories in economics *J. Evol. Econ.* **4** 153–72

Nelson R R and Winter S G 1982 *An Evolutionary Theory of Economic Change* (Cambridge, MA: : Harvard University Press)

Nowak M A 2006 *Evolutionary Dynamics: Exploring the Equations of Life* (Cambridge, MA: Belknap Press)

Nyborg K 2000 Homo economicus and homo politicus: interpretation and aggregation of environmental values *J. Econ. Behav. Organ.* **42** 305–22

Quetelet A 1835 *Sur l'homme et le Développement de ses Facultés, ou Essai de Physique Sociale [Essay on Social Physics: Man and the Development of his Faculties]* (Paris: Imprimeur-Libraire) pp 1–2 (in French)

Rosser J B Jr 2010 Is a transdisciplinary perspective on economic complexity possible? *J. Econ. Behav. Organ.* **75** 3–11

Safarzynska K and van den Bergh J C 2008 Evolutionary modelling in economics: a survey of methods and building blocks *Papers on Economics and Evolution, No. 0806* (Jena: Max Planck Institute of Economics)

Saint-Simon C H 1803 *Lettres d'un Habitant de Geneve à ses contemporains*

Schumpeter J A 1911 *The Theory of Economic Development: An Inquiry into Profits, Capital, Credit, Interest, and the Business Cycle* (New Brunswick, NJ: Transaction Books) Translated from the German by Redvers Opie (1983)

Schumpeter J A 1942 *Capitalism, Socialism and Democracy* (New York, NY: Harper & Brothers)

Schuster P and Sigmund K 1983 Replicator dynamics *J. Theor. Biol.* **100** 533–8

Silverberg G, Dosi G and Orsenigo L 1988 Innovation, diversity, and diffusion: a self-organizing model *Econ. J.* **98** 1032–54

Smith A 1776 *An Inquiry into the Nature and Causes of the Wealth of Nations* https://books.google.com/books?id=SwFYIf_E1CIC&lpg=PA31&pg=PA31#v=onepage&q&f=false

Taylor P D and Jonker L 1978 Evolutionarily stable strategies and game dynamics *Math. Biosci.* **40** 145–56

Thompson J N 1999 Specific hypotheses on the geographic mosaic of coevolution *Am. Nat.* **153S** 1–14

Veblen T 1898 Why is economics not an evolutionary science? *Q. J. Econ.* **12** 373–97
Veblen T 1961 *The Place of Science in Modern Civilization* (New York: Russell & Russell)
Veblen T 1964 *The Instinct of Workmanship* (New York: Augustus M. Kelly Booksellers)
Walras L 1874 *Elements of Pure Economics* (New York: Routledge) [1874–77] Translated from the French by Routledge (2003)
Weintraub E R 2007 Neoclassical economics *The Concise Encyclopedia of Economics* http://www.econlib.org/library/Enc1/NeoclassicalEconomics.html (retrieved 29 July 2021)
Whitham T G *et al* 2003 Community and ecosystem genetics: A consequence of the extended phenotype *Ecology* **84** 559–73
Wunder T A 2007 Toward an evolutionary economics: the 'Theory of the Individual' in Thorstein Veblen and Joseph Schumpeter *J. Econ. Issues* **41** 827–39

IOP Publishing

Forecasting with Maximum Entropy

The interface between physics, biology, economics and information theory

Hugo Fort

Chapter 5

Inferring effective interaction matrices through MaxEnt

'From the earliest times this process of plausible reasoning preceding decisions has been recognized. Herodotus, in about 500 BC, discusses the policy decisions of the Persian kings. He notes that a decision was wise, even though it led to disastrous consequences, if the evidence at hand indicated it as the best one to make; and that a decision was foolish, even though it led to the happiest possible consequences, if it was unreasonable to expect those consequences.'

—E T Jaynes (1985)

- We start this chapter by noting that the recipe to estimate the parameters of the linear Lotka–Volterra generalized equations (LLVGE) from monoculture and biculture experiments may not be applicable in more complex communities, like natural ecosystems, let alone communities of firms in markets.
- We therefore have to devise alternative procedures to estimate model parameters by more indirect methods. One possibility is to resort to the maximum entropy (MaxEnt) principle we introduced in chapter 1 as a method of making predictions from limited data which introduces minimal bias.
- As we have seen in chapters 1 and 2, the basis of the MaxEnt method is to work with known constraints on the system written as average values. Thus, in order to obtain a good candidate for the interaction matrix through MaxEnt, a key element is choosing the right constraints. In particular, the choice of the covariance matrix, which connects pairs of species, as one of the constraints is crucial to obtain a set of pairwise interaction coefficients.
- The models that result when we maximize the Shannon entropy imposing that the means and covariances of entities (species, genotypes, firms, etc) are known as *pairwise maximum-entropy* (PME) models. We comment on diverse examples in which such PME models have been used to analyze data associated with biological problems in ecology and evolution.

doi:10.1088/978-0-7503-3931-5ch5

- We conclude this chapter with some words of caution; PME models may be insufficient to model problems in neuronal systems or protein folding of long proteins. In these applications higher-order correlations may be required.

5.1 Working with imperfect information

We have seen in chapter 3 that the LLVGE serve to describe the dynamics of a community of S interacting species. Then in chapter 4 we learned how these equations can be used to model communities different than ecosystems, like markets of firms. The LLVGE involve three sets of parameters:

I. the intrinsic per-capita growth rates of species, $\{r_i\}$ (S parameters);
II. the species carrying capacities, $\{K_i\}$ (S parameters);
III. the interspecific interaction coefficients, $\{\alpha_{ij}\}$ ($S \times (S - 1)$ parameters).

In section 3.2.1 we also have shown that one way to estimate the sets (II) and (III) is by performing S monoculture experiments (isolating each species i from the rest) and $S \times (S - 1)/2$ biculture experiments (isolating each possible pair of species i and j from the rest) provided in each experiment the involved species have reached their equilibrium yields. Then the parameters are obtained through equation (3.15). To get the $\{r_i\}$ parameters we need for each species its 'species trajectory', i.e., how the yield of this species changes with time.

However, these experiments, which are common practice in agricultural science for analyzing artificial communities of species (e.g., crop mixtures), may not be feasible in more complex communities, like natural ecosystems, let alone communities of companies in markets. For example, the number of coexisting species S can be of the order of hundreds in communities like plants in natural grasslands or trees in tropical forests. Thus performing the required $\sim S^2$ experiments for estimating the LVGE parameters becomes practically impossible. Furthermore, in the case of trees in tropical forests, to create species monocultures (or bicultures) is not an option. Another limitation occurs when species are obligate mutualists, like plants and pollinators, so that these species are not viable when separated. Likewise, we cannot isolate companies from the rest of the market to study their dynamics under controlled conditions.

Therefore, we have to devise alternative procedures to estimate model parameters by more indirect methods. In the subsequent sections of this chapter we will show that one possibility is to resort to the maximum entropy (MaxEnt) method we introduced in chapter 1 as a method of making predictions from limited data which introduces minimal bias.

5.2 The Lotka–Volterra maximum entropy interaction matrix

5.2.1 Choosing the right constraints

Our goal is to derive a candidate for the interaction matrix between pairs of entities (species, firms, etc) through MaxEnt. Given that MaxEnt produces a probability distribution $\mathcal{P}(s)$ for the possible microscopic states s of a given system, our departing point will be the pair of equations (1.15) and (1.16) for $\mathcal{P}(s)$ obtained via MaxEnt. That is, assuming that the averages, $\bar{A}_i$ ($i = 1,2,\ldots$), of C quantities A_i are known, this probability distribution can be written as:

$$\mathbf{P}(s) = \frac{e^{-\sum_{i=1}^{C}\lambda_i A_i(s)}}{\sum_{s=1} e^{-\sum_{i=1}^{C}\lambda_i A_i(s)}}. \tag{5.1}$$

As mentioned, the denominator is a normalizing constant $Z(\lambda_1,\ldots,\lambda_C)$, called the *partition function*, which holds all the relevant information of the system. The Lagrange multiplier parameters λ_i are determined from the solution of the C nonlinear equations:

$$\bar{A}_i = -\frac{\partial \ln Z}{\partial \lambda_i}. \tag{5.2}$$

Let us describe a microscopic state s by a set of continuum variables $\{v_i\}$, $i = 1,\ldots,S$, which can correspond to species abundances in the case of an ecosystem, or firm market values for markets, etc. We then arrange them into a vector $\mathbf{v} = [v_1, v_2, \ldots, v_S]^{\mathrm{T}}$[1]. For clarity of notation let us denote expected values of quantities X, by E[X]. For example, the normalization condition and the entropy become, respectively:

$$E[1] \equiv \int \mathcal{P}(\mathbf{v})\mathrm{d}\mathbf{v} = \int \prod_i \mathrm{d}v_i \mathcal{P}(\mathbf{v}) = 1, \tag{5.3}$$

$$-\mathrm{E}[\ln P(\mathbf{v})] \equiv -\int d\mathbf{v}\mathcal{P}(\mathbf{v}) \ln P(\mathbf{v}) = -\int \prod_i dv_i \mathcal{P}(\mathbf{v}) \ln \mathcal{P}(\mathbf{v}) = H. \tag{5.4}$$

At the end of chapter 1 we found that if the constraint obeyed by $\mathcal{P}$ was that we know the value of the mean μ of a single real variable x, then $\mathcal{P}(x)$ was proportional to $e^{-(x/\mu)}$. Likewise, if we also know the variance we obtain the normal distribution.

Here we will extend those results we got for a single real variable x to the set of variables $v_1,v_2,\ldots,v_S$. Then we consider the situation in which we have for each variable $v_1, v_2,\ldots,v_S$ a sample of T measurements (that can correspond to different times or sites or experimental replicas, etc) from which we can compute whatever sample averages. The simplest of such averages we can include are the sample means. Since we are interested in the interactions between entities the simplest constraint we can consider that simultaneously takes into account their pairwise interactions and can be expressed as an average is the covariance. So we will also include the sample covariances. The sample mean vector and the sample covariance matrix are unbiased estimates of the mean and the covariance matrix of the random vector $\mathbf{V}$. That is, we consider that their first two statistical moments match the corresponding sample means μ_i and covariances $\mathbf{\Sigma}_{ij}$ over a sample of these

[1] The T denotes 'transpose', i.e., it transforms the row vector $[v_1, v_2,\ldots,v_S]$ into a column vector.

T measurements. These constraints on the distribution are then expressed as expected values as:

$$\mathrm{E}[v_i] = \mu_i \equiv \overline{v_i} = \sum_{t=1}^{T} v_i(t)/T, \tag{5.5a}$$

$$\begin{aligned}\mathrm{E}[(v_i - \mathrm{E}[v_i])(v_j - \mathrm{E}[v_j])] &= \Sigma_{ij} \equiv \overline{v_i v_j} - \overline{v_i}\,\overline{v_j} \\ &= \sum_{t=1}^{T} v_i(t)v_j(t)/T - \sum_{t=1}^{T} v_i(t) \sum_{s=1}^{T} v_j(s)/T^2.\end{aligned} \tag{5.5b}$$

In equations (5.5), the average of $x(t)$ is denoted with an upper bar, i.e., $\overline{x} \equiv \sum_{t=1}^{T} x(t)/T$.

Therefore, we can write the partition function of equation(5.1) as:

$$Z(\mathbf{h}, \mathbf{M}) = \int \prod_i \mathrm{d}v_i \mathrm{e}^{\sum_{i=1}^{S} h_i v_i + \frac{1}{2}\sum_{ij} M_{ij} v_i v_j}, \tag{5.6}$$

where h_i are the elements of a vector $\mathbf{h}$ of Lagrange multipliers and M_{ij} the elements of a matrix $\mathbf{M}$ of Lagrange multipliers corresponding, respectively, to constraints (5.5a) and (5.5b).

5.2.2 The MaxEnt interaction matrix and its properties

Equation (5.6) is a Gaussian integral, whose result is given by:

$$Z = \frac{(2\pi)^{S/2}}{\sqrt{\det \mathbf{M}}} \mathrm{e}^{1/2 \sum_{ij} M^{-1}{}_{ij} h_i h_j} = \frac{(2\pi)^{S/2}}{\sqrt{\det \mathbf{M}}} \mathrm{e}^{1/2 \mathbf{h}^{\mathrm{T}} \mathbf{M}^{-1} \mathbf{h}}. \tag{5.7}$$

Now we will show how to relate the unknown vector $\mathbf{h}$ and matrix $\mathbf{M}$ to the known vector $\boldsymbol{\mu}$, with elements $\overline{v_i}$, and the known covariance matrix $\boldsymbol{\Sigma}$, whose elements are defined by $\Sigma_{ij} \equiv \overline{v_i v_j} - \overline{v_i}\,\overline{v_j}$. This allows one to infer model parameters h_i and M_{ij} from empirical observations such as the means and covariances of the variables v_i. These relationships can be conveniently obtained from the derivatives of the partition function equation (5.2), which is the standard approach in statistical physics. That is,

$$\mu_k \equiv \overline{v_k} = \frac{\partial \ln Z}{\partial h_k}, \tag{5.7a}$$

$$\Sigma_{ij} \equiv \overline{v_i v_j} - \overline{v_i}\,\overline{v_j} = \frac{\partial^2 \ln Z}{\partial h_i \partial h_j}. \tag{5.7b}$$

Substituting equation (5.6) into equations (5.7a) and (5.7b) we get:

$$\boldsymbol{\mu} = \mathbf{M}^{-1}\mathbf{h}, \tag{5.8a}$$

$$\mathbf{\Sigma} = -\mathbf{M}^{-1}, \tag{5.8b}$$

which can be inverted to give:

$$\mathbf{h} = -\mathbf{\Sigma}^{-1}\boldsymbol{\mu}, \tag{5.9a}$$

$$\boxed{\mathbf{M} = -\mathbf{\Sigma}^{-1}.} \tag{5.9b}$$

Let us pause and notice that equation (5.6) has the same form as the partition function of the Ising model equation (2.40) used to describe magnetic systems. If we take a constant $M_{ij} = J$, and a constant $h_i = h$, we can identify the matrix of Lagrange multipliers with a coefficient measuring the strength of the interaction between spins, while the vector h corresponds to an external magnetic field.

Therefore, the matrix $\mathbf{M}$ has the natural interpretation of an *effective* interaction matrix between entities. Equation (5.9b) is a very important result; it tells us that this interaction can be obtained as minus the inverse of the covariance matrix. The resulting models are called ***pairwise maximum-entropy*** (PME) ***models*** (Stein *et al* 2015). We stress the effective character of these interaction coefficients. This is because firstly the covariances were computed not from pairs of isolated entities (species, companies, etc) under controlled conditions but from pairs of entities immersed in a community. Thus they embody all the multiple interactions among the entities as well as interactions with 'the environment' in which they live. This implies that in many cases, PME models can correctly reproduce correlations of order higher than two which were included as constraints (see, e.g., Lezon *et al* 2006, Schneidman *et al* 2006 for examples). On top of this, the inverse of an $S \times S$ matrix involves products of S of their elements, which indirectly connect each entity with all the others (see Box 5.1).

Notice that the choice of the covariance matrix, which connects pairs of species, as a required constraint was crucial to obtain a set of pairwise interaction coefficients.

We conclude this section with two properties of matrix $\mathbf{M}$:

a. Since the covariance matrix $\mathbf{\Sigma}$ has positive elements along the diagonal, the minus sign in the rhs of equation (5.9b) implies that the diagonal elements of $\mathbf{M}$—corresponding to the **self-interaction coefficients—must be negative**.
b. Since $\mathbf{\Sigma}$ is symmetric, $\mathbf{M}$ is also symmetric.

These two properties follow from equation (vi) of Box 5.1. If we combine this equation with equation (5.9b) we get $M_{ii} = -1/\lambda_i$, where is the ith eigenvalue. Since the covariance matrix is a positive semi-definite matrix[2], this proves property (a). The symmetry of $\mathbf{M}$ is immediate from equation (vi) of Box 5.1. Indeed, this symmetry poses a limitation for using $\mathbf{M}$ as an interaction matrix because in real situations the strength and sign of the effect of an entity i over j is in general not

[2] Covariance matrix. Encyclopedia of Mathematics. URL: http://encyclopediaofmath.org/index.php?title=Covariance_matrix&oldid=46540

Box 5.1. Inverse of a matrix in terms of its adjugate

Definitions:

- The *(i,j)-minor* of a matrix **A**, denoted $\mathcal{M}_{i,j}$, is the determinant of the $(n-1) \times (n-1)$ matrix that results from deleting row i and column j of **A**.
- The *cofactor matrix* of **A** is the $n \times n$ matrix C_A whose (i,j) entry is the (i,j) cofactor of **A**, which is the (i,j)-minor times a sign factor:

$$\mathbf{C_A} = \left[(-1)^{i+j}\mathcal{M}_{i,j}\right]_{1\leqslant i,\, j\leqslant n} \tag{i}$$

- The *adjugate* of A is the transpose of the cofactor matrix **C**, defined as:

$$\mathrm{adj}(\mathbf{A}) = \mathbf{C}_\mathbf{A}^\mathrm{T} = [(-1)^{i+j}\mathcal{M}_{j,\, i}]_{1\leq i,\, j\leq n} \tag{ii}$$

A property of the determinant of a matrix A is that it can be expressed as the sum of each element of a line (row or column) by its cofactor, i.e.,

$$\det(\mathbf{A}) = \sum_{j=1}^{n} a_{ij}(-1)^{i+j}\mathcal{M}_{i,j} = \sum_{j=1}^{n} a_{ji}(-1)^{i+j}\mathcal{M}_{j,\, i}. \tag{iii}$$

Another property is that the sum of each element of a line (row or column) by the cofactor of other line is 0, i.e.,

$$\sum_{j=1}^{n} a_{ij}(-1)^{k+j}\mathcal{M}_{k,j} = 0 = \sum_{j=1}^{n} a_{ji}(-1)^{k+j}\mathcal{M}_{j,\, k} \quad \text{if} \quad i \neq k. \tag{iv}$$

From (iii) and (iv) imply that the product of A with its adjugate yields a diagonal matrix whose diagonal entries are the determinant $\det(A)$. That is,

$$\mathbf{A}^{-1} = \mathrm{adj}(\mathbf{A})/\det(\mathbf{A}). \tag{v}$$

where I is the $n\times n$ identity matrix.

Therefore, the above formula implies that

$$\boxed{\mathbf{A}^{-1} = \mathrm{adj}(\mathbf{A})/\det(\mathbf{A}).} \tag{vi}$$

equal to the one of j over i. In the next subsection we will see how this drawback can be overcome.

5.2.3 A non-symmetrical MaxEnt interaction matrix

According to table 3.1, a Lotka–Volterra generalized matrix is a non-dimensional matrix, in which it is customary to take its diagonal elements equal to −1. We can obtain such an interaction matrix **I** from the MaxEnt matrix **M**, which by equation (5.9b) has dimensions of v_i^2, by dividing the entries of each row i of **M** by the corresponding absolute value of the diagonal element M_{ii}:

$$I_{ij} = \frac{M_{ij}}{|M_{ii}|} \quad i,j = 1,\ldots,S. \tag{5.10}$$

Additionally, notice that matrix **I** is no longer symmetric. In this sense **I** represents an improvement over **M** since it allows one to model asymmetric interactions.

The MaxEnt normalized matrix equation (5.10) will be used to approach the dynamics of real data temporal series in the next two applications, each considered in a separate chapter. Chapter 6 is devoted to forecasting the trajectories of tree species in tropical forests and finding early warnings of species crashes. In chapter 7 we will analyze the dynamics of markets, both qualitatively and quantitatively.

5.3 How good is the pairwise approximation?

In the last 15 years, PME probability models have been used to analyze biological data associated to diverse problems, such as gene network inference (Lezon *et al* 2006, Locasale and Wolf-Yadlin 2009), analysis of neural populations (Schneidman *et al* 2006, Tang *et al* 2008), protein contact prediction (Weigt *et al* 2009, Marks *et al* 2011, Jones *et al* 2012), modelling of animal flocks (Bialek *et al* 2012), and community ecology (Volkov *et al* 2009).

Actually, inferring the values of the interaction coefficients underlying biological systems is crucial to develop applications, such as protein design or multi-gene effects in relating variants to phenotypic changes as well as multi-genic traits (Rockman 2008, Ritchie *et al* 2015). Networks of neurons provide another interesting application. If we look in a small window of time, then each neuron either does or does not generate an action potential, or spike (Bialek and Ranganathan 2008). It has been suggested that the full pattern of correlations among all the neurons in such a network can be described by the MaxEnt model that is consistent with the observed pairwise correlations (Schneidman *et al* 2003). This approach has been shown to provide successful predictions for the combinatorial patterns of activity in the vertebrate retina (Schneidman *et al* 2006). These maximum entropy models in fact are instances of our old friend, the Ising model, which has long been discussed as a schematic model for neural networks (Hopfield 1982, Amit 1989).

However, a word of caution regarding pairwise modelling is in order. Although it is true that pairwise interactions can provide considerable explanatory power, restricting our description to pairwise interactions is an enormous simplification. In the case of biological systems often they have many elements (e.g., bases along DNA, amino acids along a single protein chain, proteins, cells, etc) with many opportunities to interact in combinatorial fashion as they generate biological function (Bialek and Ranganathan 2008). It turns out that PME models may be insufficient to model problems like neuronal spike trains for systems involving many neurons or protein folding of proteins longer than just a few amino acids (Roudi *et al* 2009).

Therefore, higher-order correlations may be required. Indeed, the partition function equation (5.6) can be considered as the lowest order approximation that takes into account correlations between entities of a general MaxEnt expansion of the form:

$$Z(\mathbf{h}, \mathbf{M}, \mathbf{N}, \ldots) = \int \prod_i \mathrm{d}v_i \mathrm{e}^{\sum_{i=1}^{S} h_i v_i + \frac{1}{2}\sum_{ij} M_{ij} v_i v_j + \sum_{ijk} N_{ijk} v_i v_j v_k + \cdots}. \tag{5.11}$$

For instance, the circumstances in which the impact of the higher-order interactions appearing in equation (5.11) is non-negligible have been studied in neural circuits in which correlations are generated by common input (Barreiro *et al* 2014). This impact is quantified as the distance of the PME probability distribution, p_{pair}, to the true probability distribution, p_{true}. A natural measure of the distance between p_{pai} and p_{true} is the Kullback–Leibler (1951) divergence. A very recent study (Chelaru *et al* 2021) reports that the model based on pairwise interactions captured ~90% of the spiking activity structure during wakefulness and sleep. However, regardless of brain state, pairwise interactions fail to explain experimentally observed collective behavior of cortical populations in executive areas. That is, explaining the population dynamics of neurons in prefrontal cortex requires high-order interactions.

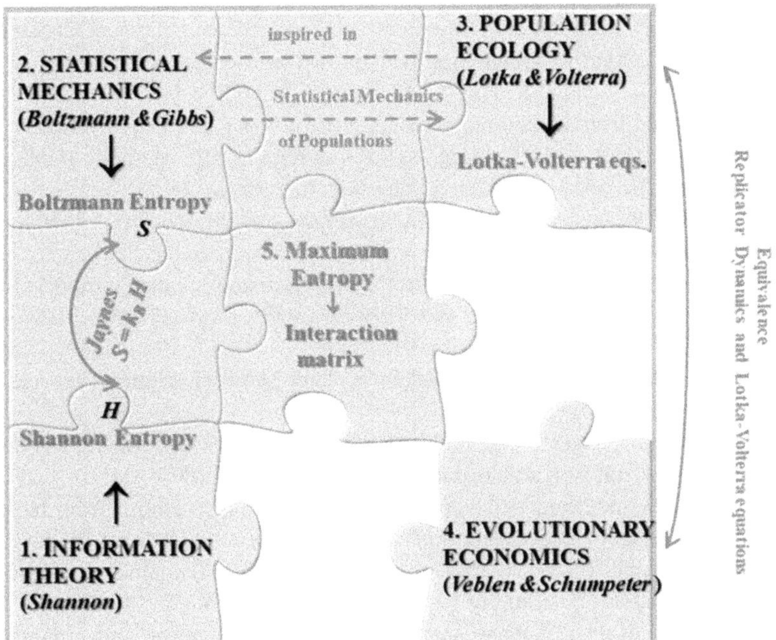

References

Amit D J 1989 *Modeling Brain Function: The World of Attractor Neural Networks* (Cambridge: Cambridge University Press)

Barreiro A K, Gjorgjieva J, Rieke F and Shea-Brown E 2014 When do microcircuits produce beyond-pairwise correlations? *Frontiers Comput. Neurosci.* **8** 1–25

Bialek W and Ranganathan R 2008 *Rediscovering the power of pairwise interactions* (arXiv:0712.4397 [q-bio.QM])

Bialek W, Cavagna A, Giardina I, Mora T, Silvestri E and Viale M *et al* 2012 Statistical mechanics for natural flocks of birds *Proc. Natl Acad. Sci.* **109** 4786–91

Chelaru M I, Eagleman S, Andrei A R, Milton R, Kharas N and Dragoi V 2021 High-order interactions explain the collective behavior of cortical populations in executive but not sensory areas *Neuron* **109** P3954–61

Hopfield J J 1982 Neural networks and physical systems with emergent collective computational abilities *Proc. Natl Acad. Sci.* **79** 2554–8

Jaynes E T 1985 Bayesian methods: general background *Maximum Entropy and Bayesian Methods in Applied Statistics* ed J H Justice (Cambridge: Cambridge University Press) pp 1–25

Jones D T, Buchan D W A, Cozzetto D and Pontil M 2012 PSICOV: precise structural contact prediction using sparse inverse covariance estimation on large multiple sequence alignments *Bioinformatics* **28** 184–90

Kullback S and Leibler R 1951 On information and sufficiency *Ann. Math. Stat.* **22** 79–86

Lezon T R, Banavar J R, Cieplak M, Maritan A and Fedoroff N V 2006 Using the principle of entropy maximization to infer genetic interaction networks from gene expression patterns *Proc. Natl. Acad. Sci.* **103** 19033–8

Locasale J W and Wolf-Yadlin A 2009 Maximum entropy reconstructions of dynamic signaling networks from quantitative proteomics data *PLoS One* **4** e6522

Marks D S, Colwell L J, Sheridan R, Hopf T A, Pagnani A and Zecchina R *et al* 2011 Protein 3D structure computed from evolutionary sequence variation *PLoS One* **6** e28766

Ritchie M D, Holzinger E R, Li R, Pendergrass S A and Kim D 2015 Methods of integrating data to uncover genotype–phenotype interactions *Nat. Rev. Genet.* **16** 85–97

Rockman M V 2008 Reverse engineering the genotype–phenotype map with natural genetic variation *Nature* **456** 738–44

Roudi Y, Nirenberg S and Latham P E 2009 Pairwise maximum entropy models for studying large biological systems: when they can work and when they can't *PLoS Comput. Biol.* **5** e1000380

Schneidman E, Still S, Berry M J and Bialek W 2003 Network information and connected correlations *Phys. Rev. Lett.* **91** 238701

Schneidman E, Berry M J, Segev R and Bialek W 2006 Weak pairwise correlations imply strongly correlated network states in a neural population *Nature* **440** 1007–12

Stein R R, Marks D S and Sander C 2015 Inferring pairwise interactions from biological data using maximum-entropy probability models *PLoS Comput. Biol.* **11** e1004182

Tang A, Jackson D, Hobbs J, Chen W and Smith J *et al* 2008 A maximum entropy model applied to spatial and temporal correlations from cortical networks *in vitro J. Neurosci.* **28** 505–18

Volkov I, Banavar J R, Hubbell S P and Maritan A 2009 Inferring species interactions in tropical forests *Proc. Natl Acad. Sci.* **106** 13854–9

Weigt M, White R A, Szurmant H, Hoch J A and Hwa T 2009 Identification of direct residue contacts in protein–protein interaction by message passing *Proc. Natl Acad. Sci.* **106** 67–72

Chapter 6

Early warning indications of species crashes from effective intraspecific interactions in tropical forests

This chapter is mainly based on Fort and Grigera (2021b).

The author acknowledges the Center for Tropical Forest Science for providing data for the BCI plot (Condit *et al* 2019).

> 'We need have no fear of making shaky calculations on inadequate knowledge; for if our predictions are indeed wrong, then we shall have an opportunity to improve that knowledge, an opportunity that would have been lost had we been too timid to make the calculations. Instead of fearing wrong predictions, we look eagerly for them; it is only when predictions based on our present knowledge fail that probability theory leads us to fundamental new knowledge.'
>
> —E T Jaynes (1985)

We start by noting a challenge for conservation biologists and agencies working to sustain the ecosystem services, namely the vulnerability of biodiversity of an ecological community to several factors like, climate change, habitat fragmentation, resource exploitation, etc. Hence, there is a clear need for early warning indicators of species loss generated from empirical data, which is the topic covered in this chapter.

With this goal in mind we use the tree community of the long-term 50 hectare plot on Barro Colorado Island (BCI), Panama, which is one of the most intensively studied in the world. This plot was established in 1981 and fully censused in 1982, and then every 5 years from 1985 through 2015 (the 2020 census remains on stand-by due to COVID-19). This extensive dataset reveals that some tree species suffered steep population declines.

Here we propose an early warning indicator of such tree population crashes and test it against the BCI dataset. This early warning indicator is based on the pairwise

doi:10.1088/978-0-7503-3931-5ch6

maximum entropy (PME) procedure (introduced in chapter 5). That is, the spatial covariance matrices, $\mathbf{\Sigma}$, of the 20 most abundant tree species in BCI allow us to compute, via the maximum entropy (MaxEnt) principle, the effective interaction matrices, M, among these species for the eight censuses available from 1982 to 2015. For each species i and each census c, the absolute value of the intraspecific competition coefficients $M_{ii}(c)$ are much larger than those of the interspecific interaction coefficients $M_{ij}(c)$ with $i \neq j$. We show that this result can be derived from a similar empirical relationship observed for the covariance matrices.

The main finding of this chapter is that for those tree species that suffered steep population declines (of at least 50%), across the eight tree censuses, the drop of M_{ii} is always steeper and occurs before the drop of the corresponding species abundance N_i. Indeed, such sharp declines in M_{ii} occur between 5 and 15 years in advance of comparable declines for N_i, and thus they serve as early warnings of impending population busts. Furthermore, this drop of M_{ii} is linked to the anomalous variance, which is a known early warning of incoming catastrophic shifts.

6.1 Background: diversity loss and early warning signals

6.1.1 On fluctuations of biodiversity and what drives these changes

The rate of species loss we are observing around the world is much greater than anything experienced historically (Thompson and Starzomski 2006). Diversity loss, combined with environmental change, increases the risk of abrupt and potentially irreversible ecosystem collapse (Ives and Carpenter 2007, Hooper *et al* 2012, MacDougall *et al* 2013). Such catastrophic ecological regime shifts may be announced in advance by early warning signals such as slowing return rates from perturbation and rising variance (Carpenter *et al* 2011). Thus, it is difficult to overstate the importance of identifying early warning signals that would allow managers to predict catastrophic biodiversity losses before they happen so that they can take remedial action.

The issue of fluctuations of biodiversity is closely related with a major debate in ecology of whether species coexistence emerges from equilibrium niche partitioning or from non-equilibrium stochastic dispersal-assembly (Clark and McLachlan 2003, Ishida *et al* 2003). According to the first hypothesis, in a community at equilibrium each species occupies a different niche that results from and reduces direct competition (Whittaker 1975). Stabilizing mechanisms—like tradeoffs between species in terms of their capacities to disperse to sites where competition is weak, to exploit abundant resources effectively and to compete for scarce resources (Clark and McLachlan 2003)—play an important role. Alternatively, the dispersal-assembly hypothesis assumes that communities are open non-equilibrium assemblages of species that coexist only transiently through chance, history, and random dispersal rather than by the stabilizing effects of niche differentiation, regarded as superfluous (Hubbell 2008, 2009). This 'neutral model' thus emphasizes 'equalizing' mechanisms (Chesson 2000), because competitive exclusion of similar species is slow. The relative importance of these two mechanisms is still a matter of discussion. It was argued that stabilizing processes and fitness inequality vary among communities and respond to anthropogenic changes (Adler *et al* 2007). However, strong evidence was found that patterns of habitat association (and hence conclusions on the importance of niche versus neutral processes) are affected by the choice of sampling scale (Garzon-Lopez *et al* 2014, Chase 2014). Indeed, Garzon-Lopez *et al* (2014) found that the very BCI tropical forest is highly niche-structured at large scales, and largely neutrally structured at small scales.

6.1.2 Early warning signals

A general problem is that our world is changing at an unprecedented rate and we need to understand the nature of the change and to forecast the way in which it might affect ecosystems of interest. Additionally, there is increasing evidence that ecosystems can pass thresholds and go through catastrophic regime shifts (Scheffer and Carpenter 2003, Boettiger *et al* 2013) where sudden and large changes in their functions take place with serious consequences for human well-being (Millennium Ecosystem Assessment 2005). Additionally, these changes are very difficult and costly to reverse. Thus, an important problem in environmental sciences is to get early warning signals of these impending catastrophic regime shifts in ecosystems to allow addressing currently intractable problems in ecosystem management, such as the avoidance of ecological surprises, and the maintenance of systems in desired states (Fort 2020). There is empirical evidence that diversity loss increases vulnerability to ecosystem collapse (MacDougall *et al* 2013). This is why devising indicators that work as early warnings of severe decreases in the abundance of species, generated from empirical data, is an important task.

In order to address questions about the impact of environmental change, and to understand what might be done to mitigate the predicted negative effects, ecologists need to develop the ability to project models into novel, future conditions. A main difficulty with the majority of current ecological models is that they are excellent at describing the way in which a system has behaved, but they are poor at predicting its future state, especially under novel conditions (Evans 2012).

In chapter 3, we have mentioned that a common simplification of these mathematical ecological models is that they rely on the well-mixed assumption or, in the physics parlance, the mean-field (MF) approximation. For example, using such a mean field, i.e., non-spatially explicit, model it was shown that the rising of the temporal variance for the nutrient concentration in lakes works as an early warning signal of a catastrophic shift towards eutrophication (Carpenter and Brock 2006).

Later on it was shown that, if one takes into account explicitly the space, the spatial variance provides an even earlier early warning (Fernández and Fort 2009, Donangelo *et al* 2010, Dakos *et al* 2010). This is an interesting motivation for moving from the MF assumption and thus resorting to spatially explicit modelling.

6.2 Goal

Our goal is to provide early warnings of sudden drastic drops in the number of individuals of species in a community (in our case, the trees of a tropical forest) before they actually occur. To do this we need to know which variable(s) to monitor.

It was recently found that intraspecific interactions between trees in BCI are much larger than the interspecific interactions (Fort 2020, Fort and Grigera 2021a, 2021b). Furthermore, self-regulation by intraspecific competition seems to control the trajectories (i.e., the corresponding sequences of abundances over censuses) of several species in the BCI plot (Fort and Grigera 2021a). This opens the possibility that the intraspecific interaction coefficients could provide early warnings for impending species crashes. Hence, firstly we will derive analytically this dominance of the intraspecific interaction coefficients by estimating them using the MaxEnt procedure of chapter 5. Next we will focus on these diagonal elements of the estimated interaction matrices, analyzing if their behavior serves to provide early warnings for species that crashed before they did it.

6.3 Data

The tree community of the long-term 50 hectare plot on Barro Colorado Island (BCI), Panama, is one of the most intensively studied in the world. This plot was established in 1981 and fully censused in 1982 then every 5 years from 1985 through 2015. All stems ⩾1 cm diameter-at breast-height (dbh) were mapped, measured, and identified in each census (Condit *et al* 2017). Figure 6.1 shows maps of distributions of trees for the eight most abundant species in the first census of 1982. Each dot represents a tree of a given species in the 50 hectare plot of BCI.

Data from eight censuses of trees spanning almost 35 years—1982, 1985, and then every five years until 2015—is a period long enough to examine decadal changes in tree growth rates and death rates. They demonstrate that the BCI forest has exhibited considerable dynamism in this period. A variety of population trajectories can be observed: busts, recoveries and oscillations (Condit *et al* 2017).

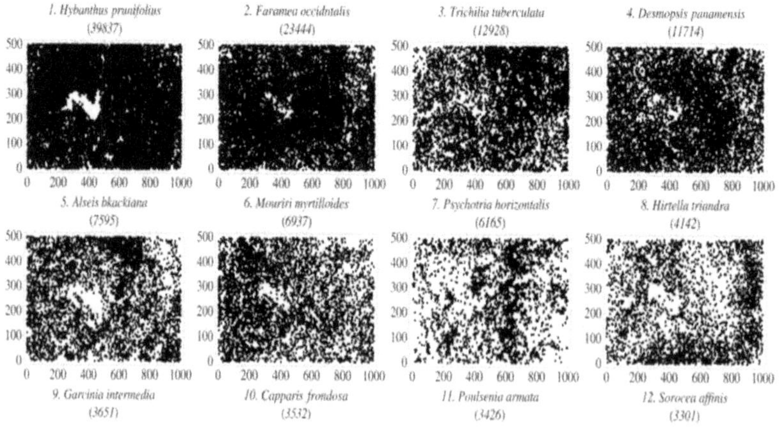

Figure 6.1. Species Distribution Maps from the 50 ha Plot at Barro Colorado Island for the eight most abundant species in the first census of 1982. Each dot represents a tree of a given species in the 50 ha plot of BCI. Reprinted from Condit (1998), copyright 1998 with permission of Springer.

6.4 Estimating the interaction matrix through MaxEnt

In our case, the system is the community of trees of the 50 hectare BCI plot. This community is composed of S coexisting species for each of which we have spatial information on the location of every individual. With the aim to use these data to infer the effective interaction strengths between the species we follow the MaxEnt principle of chapter 5. The idea is to obtain the effective interaction strengths between tree species from the expression of the probability distribution that maximizes the information entropy with constraints. Indeed, using MaxEnt, Volkov *et al* (2009) estimated the interaction strengths between species for the 20 most abundant species for trees (dbh $\geqslant$ 1 cm) in the 50 hectare BCI plot at the first census carried out in 1982.

Here we will repeat the procedure of Volkov *et al* (2009) for all the eight available censuses. That is, to compute covariances from spatial distributions of trees, we divide the 50 hectare plot into equally sized quadrats large enough to contain many individuals yet small enough to have a sufficient number of quadrats to facilitate statistical averaging. Quadrats of 20 m a side serve to solve this trade-off. Therefore, we have a number of quadrats $Q = 500\ 000\ \mathrm{m}^2/400\ \mathrm{m}^2 = 1250$. We denote each quadrat by a super index μ. The state of this quadrat is specified by the set of the densities for each species within this quadrat $\{A + B\}$, where $A + B$ is the density of the ith species (a real variable) in this particular quadrat.

The covariance matrix for census c (c runs from 1 to 8 since there are eight available censuses) can thus be computed as:

$$\Sigma_{ij}(c) = \overline{x_i(c)x_j(c)} - \overline{x_i(c)}\ \overline{x_j(c)} = \sum_{\mu=1}^{Q} x(c)_i^{\mu} x(c)_j^{\mu}/Q - \sum_{\mu=1}^{Q} x_i^{\mu}(c) \sum_{\nu=1}^{Q} x(c)_j^{\nu}/Q^2, \quad (6.1)$$

where a bar over a quantity denotes its sample average and the indices μ and ν denote quadrats The interaction matrix for census c is obtained by equation (5.9) as

$$\mathbf{M}(c) = -\mathbf{\Sigma}^{-1}(c). \quad (6.2)$$

Hence, the magnitude of M_{ij} corresponds to the strength of the net interaction of species j over species i and that the sign of M_{ij} corresponds to whether this effect is positive or negative. Notice that, since the covariance matrix $\mathbf{\Sigma}$ has positive elements along the diagonal, the minus sign in the rhs of equation (6.2) implies that the diagonal elements of M—corresponding to the intraspecific interaction coefficients—must be negative and thus correspond to intraspecific competition.

If we perform the calculations of equation (6.1), for the eight available censuses we observe these two facts about the interspecific elements of the covariance matrix (Fort and Grigera 2021a, 2021b):

(i) They take both positive and negative values, corresponding, respectively, to positive and negative correlations (the intraspecific elements $\Sigma_{ii} = \overline{(x_i - \overline{x_i})^2}$, equal to variances σ_i^2, are by definition always positive).

(ii) They are much smaller than the intraspecific ones, by approximately an order of magnitude when taking their absolute values (table 6.1).

Table 6.1. Mean values of interspecific over intraspecific matrix elements for the eight censuses (dbh > 10 mm). In the third column the ratios $\langle|M_{ij}|\rangle/\langle|M_{ii}|\rangle$ are computed from $M = \Sigma^{-1}$; the last column is the formula of equation (6.6).

Census #	$\varepsilon \equiv \langle|\Sigma_{ij}|\rangle/\langle\Sigma_{ii}\rangle$	$\langle|M_{ij}|\rangle/\langle|M_{ii}|\rangle$	$\varepsilon/(1+18\varepsilon)$
1	0.058	0.054	0.028
2	0.058	0.055	0.028
3	0.054	0.055	0.027
4	0.050	0.046	0.026
5	0.048	0.019	0.026
6	0.046	0.013	0.025
7	0.046	0.015	0.025
8	0.047	0.017	0.026

We will present a heuristic argument, in terms of averages over the S species, to show that (ii) implies that a similar relationship holds for the interspecific interaction coefficients compared with the intraspecific ones. Indeed, this finding is in agreement with empirical evidences supporting that pairwise interspecific interaction strengths are often much weaker than the intraspecific ones from plants (Adler *et al* 2018) and across several taxonomic groups (Fort and Segura 2018). Larger intraspecific competition relative to interspecific competition is also a result expected from theoretical grounds, to ensure the stable coexistence of multispecies communities (Chesson 2000).

Hence, let us denote averages over the S species by angle brackets '$\langle\rangle$' and approximate, for each census, the covariance matrix with matrix $\mathbf{\Sigma}^{\mathrm{av}}$ in which all the diagonal elements are equal to $\langle\Sigma_{ii}\rangle$ (i.e., the average over the species variances) and all off-diagonal elements equal to $\langle|\Sigma_{ii}|\rangle$ (i.e., the average of the absolute value of the off-diagonal species covariances). The absolute value for interspecific elements is taken by (i), to avoid cancelations of interspecific covariances with opposite signs when computing averages over species. Therefore, $\mathbf{\Sigma}^{\mathrm{av}}$ can be written as:

$$\mathbf{\Sigma}^{\mathrm{av}} = \langle\Sigma_{ii}\rangle\begin{pmatrix} 1 & & \langle|\Sigma_{ij}|\rangle/\langle\Sigma_{ii}\rangle \\ & \ddots & \\ \langle|\Sigma_{ij}|\rangle/\langle\Sigma_{ii}\rangle & & \ddots \\ & & & 1 \end{pmatrix} = \langle\sigma_i^2\rangle\begin{pmatrix} 1 & & \\ & \ddots & \varepsilon \\ & \varepsilon & \ddots \\ & & & 1 \end{pmatrix}, \tag{6.3}$$

where $\varepsilon \equiv \langle|\Sigma_{ij}|\rangle/\langle\Sigma_{ii}\rangle$. It is easy to show that the inverse of matrix with ones along the principal diagonal and all non-diagonal elements equal to a constant, like that of equation (6.3), produces a matrix proportional to it. Thus, taking minus the inverse of Σ^{av} we obtain a simple expression for the effective interaction matrix M^{av} matrix, given by:

$$\boldsymbol{M}^{av} = \frac{1}{\langle\sigma_i^2\rangle} \frac{1 + (S-2)\varepsilon}{1 + (S-2)\varepsilon - (S-1)\varepsilon^2} \begin{pmatrix} -1 & & \frac{-\varepsilon}{1+(S-2)\varepsilon} \\ & \ddots & \\ \frac{-\varepsilon}{1+(S-2)\varepsilon} & & \ddots \\ & & & -1 \end{pmatrix}. \tag{6.4}$$

Thus,

$$\frac{\left| M_{ij}^{av} \right|}{\left| M_{ii}^{av} \right|} = \frac{\varepsilon}{1 + (S-2)\varepsilon} \tag{6.5}$$

For $S = 20$ (e.g., the 20 most abundant species), we have:

$$\left| M_{ij}^{av} \right| / \left| M_{ii}^{av} \right| = \frac{\varepsilon}{1 + 18\varepsilon}, \tag{6.6}$$

and ε varies between 0.046 and 0.058 across censuses, as shown in table 6.1. Therefore $1 + 18\varepsilon$ is always ≈ 2 and thus the ratio $\langle|M_{ij}|\rangle/\langle|M_{ii}|\rangle$, computed from $M = \Sigma^{-1}$, is always (smaller but) comparable to $\varepsilon \equiv \langle|\Sigma_{ij}|\rangle/\langle\Sigma_{ii}\rangle$.[1]

So we can conclude that the empirical result that the variances of the species abundances are much larger than the covariances between the abundances of different species implies, at least on average, that the intraspecific effective interaction coefficients are also much larger than the interspecific ones.

This suggests that keeping only intraspecific competition would be enough to predict the overall dynamics of the abundances of tree species. Actually, it was shown that this is the case. In the next section we sketch how this outcome is obtained; we refer the interested reader for details to Fort and Grigera (2021a).

6.5 Intraspecific competition interactions are enough to predict the trajectories of tree species

Our starting point is the linear Lotka–Volterra generalized equations (LLVGE) governing the dynamics for S interacting species within a single trophic level, equation (3.13). If we call $N_i(c)$ the abundance of species i at census number c, then the finite difference version of these equations can be written in terms of the densities (abundances divided by the plot area) $x_i(c) = N_i(c)/50000$ as

$$x(c+1)i - x_i(c) = r_i x_i(c) \left(1 + \frac{\sum_{j=1}^{S} \alpha_{ij}(c) x_j(c)}{K_i} \right) \tag{6.7}$$

[1] If we consider the total number of species in the plot (which ranged across the eight censuses from 290 to 320 species) the ratio between interspecific and intraspecific interaction coefficients becomes even smaller.

There are three sets of parameters: the intrinsic growth rates of each species, $\{r_i\}$, their carrying capacities, $\{K_i\}$, and the interaction coefficients between species, $\{\alpha_{ij}\}$.

Hence, equating $\alpha_{ij}(c)$ to $M_{ij}(c)$, we have the interaction coefficients. On the other hand, to estimate parameters r_i and K_i of the LLVGE, as is customary in time series forecasting analysis, we partition the series of observed densities, $x_i(c)$, into an earlier training period ($c = 1, 2,\ldots,c_{tr}$) and a later validation period ($c = t_{tr}+ 1,\ldots,8$). The training period is used to estimate these parameters (Shmueli and Lichtendahl 2016).

Once we have obtained the model parameters of the LLVGE from the training set of data, to assess their predictive performance on new data we can compare the abundances generated by the model against the empirical abundances of the validation set. That is, starting with $x_i^{\mathrm{p}}(c_{tr}) = x_i(c_{tr})$ (the superscript p is used to distinguish predicted variables from empirical ones) the subsequent theoretical densities are generated via equation (6.7) as:

$$x^{p}i(c+1) = x_i^{p}(c) + r_i x_i^{p}(c)\left(1 + \frac{\sum_{j=1}^{S} M_{ij}(c) x_j^{p}(c)}{K_i}\right), \quad c = c_{tr},\ldots,8. \qquad (6.7')$$

When ‘turning off’ the interspecific interactions (i.e., taking $M_{ij}= 0$), the Lotka–Volterra equations, equation (6.7), reduce to the logistic equation, equation (3.3), and thus the Lotka–Volterra predictor simplifies to the *Logistic predictor*:

$$x^{p}i(c+1) = x_i^{p}(c) + r_i x_i^{p}(c)\left(1 - \frac{M_{ii}(c) x_i^{p}(c)}{K_i}\right), \quad c = c_{tr},\ldots,8. \qquad (6.8)$$

Figure 6.2 shows comparisons of predicted abundances by equation (6.8), $N_i^{\mathrm{p}}(c) = x_i^{\mathrm{p}}(c) \times 50\,000$, against empirical abundances for eight selected species when using $c_{tr} = 6$. We can see that for most of the species the logistic equation, equation(6.8), does a pretty good job (although it fails for predicting the future abundances of some particular species, like *Cecropia insignis* (panel g)).

In summary: the conclusion of section 6.4 that $|\Sigma_{ij}|/|\Sigma_{ii}| \ll 1$ implies that $|M_{ij}|/|M_{ii}| \ll 1$, together with the result that keeping only intraspecific competition is enough to predict the overall evolution of the abundances of tree species with accuracy, lead us to focus on analyzing the dynamics of the intraspecific interaction coefficients along censuses to anticipate drastic changes in species abundances. Thus, in the next section, we compare $M_{ii}(c)$ against the experimental abundance trajectories $N_i(c)$.

6.6 A new early warning signal

The abundance trajectories $N_i(c)$ varied from monotonic growth to oscillations and busts, as shown in figure 6.3(a) for the 20 most abundant species. (To facilitate visualization the abundances are normalized by dividing with respect to the value at the first census (1982).) These abundances of these species are listed in table 6.2 for the eight censuses that were carried out until 2015.

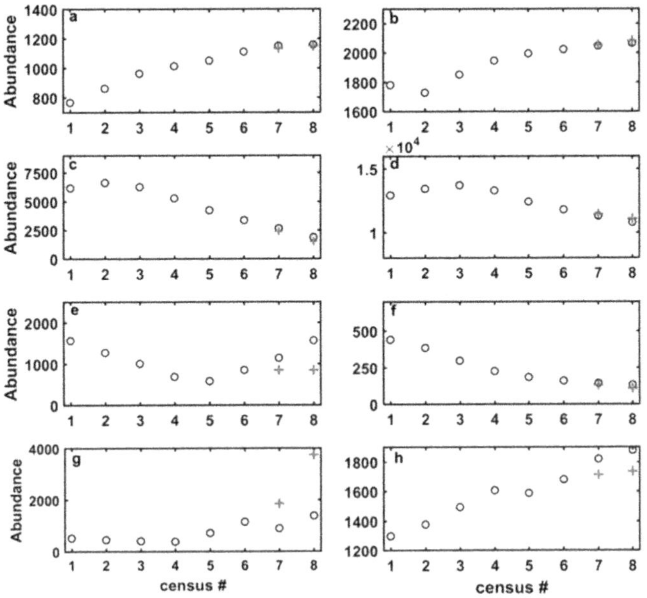

Figure 6.2. Empirical abundances of eight selected species (dbh > 1 cm) along the eight censuses (black circles) and predictions for census #7 and #8 (years 2010 and 2015) using six censuses for training, i.e., $c_{tr} = 6$ (red '+' symbols). (a) *Cassipourea elliptica*, (b) *Oenocarpus mapora*, (c) *Psychotria horizontalis*, (d) *Trichilia tuberculata*, (e) *Acalypha diversifolia*, (f) *Olmedia aspera*, (g) *Cecropia insignis*, and (h) *Tabernaemontana arborea.*

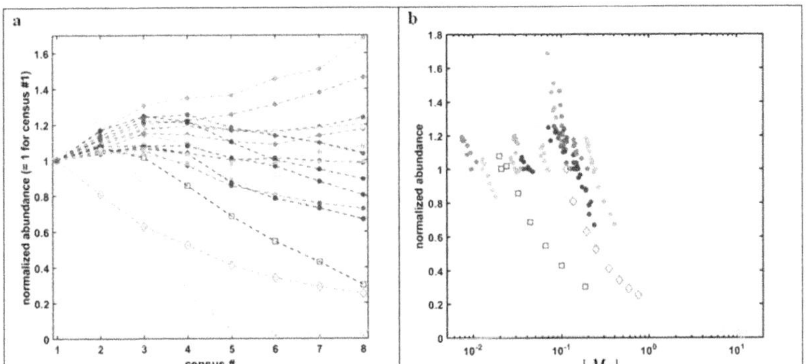

Figure 6.3. Variation of the abundances of the 20 most abundant species and of their corresponding intraspecific competition coefficients for dbh ⩾ 10 mm along the eight censuses. The abundances are normalized by dividing with respect to the value at the first census (1982). The three species which experimented population busts are indicated with open symbols (the remaining 17 species are represented by filled dots). Species # 7 *Psychotria horizontalis* (diamonds), species # 10 *Piper cordulatum* (triangles) and *Poulsenia armata* species # 19 (squares). (a) Normalized abundances versus census number. (b) Normalized abundances versus absolute value of the intraspecific competition coefficient (log scale) along different censuses. Reprinted from Fort and Grigera (2021b), copyright (2021), with permission from Elsevier.

Table 6.2. Abundances of the 20 most abundant species in census # 2 in the BCI 50 ha plot for census # 1 to # 8.

sp #		c #1	c #2	c #3	c #4	c #5	c #6	c #7	c #8
1	*Hybanthus prunifolius*	39825	42041	42055	38532	34038	32027	30117	28960
2	*Faramea occidentalis*	23442	25464	27475	27931	27478	26757	25732	24449
3	*Trichilia tuberculata*	12925	13435	13705	13294	12413	11782	11283	10781
4	*Desmopsis panamensis*	11707	12362	12514	12191	11714	11839	11637	11505
5	*Alseis blackiana*	7599	8328	8967	9046	8818	8763	8971	9113
6	*Mouriri myrtilloides*	6924	7788	7895	7581	6863	6938	7236	7459
7	*Psychotria horizontalis*	6159	6620	6250	5270	4222	3350	2641	1856
8	*Hirtella triandra*	4143	4720	5137	5198	4902	4708	4548	4289
9	*Garcinia intermedia*	3649	4064	4387	4459	4584	4790	5034	5340
10	*Piper cordulatum*	3142	3898	2678	787	159	93	87	98
11	*Capparis frondosa*	3536	3823	3823	3822	3581	3410	3107	2840
12	*Tetragastris panamensis*	3281	3816	4288	4424	4473	4777	4951	5524
13	*Sorocea affinis*	3300	3453	3407	3239	2937	2664	2405	2214
14	*Tachigali versicolor*	2922	3028	3168	3027	2547	2291	2133	1950
15	*Protium tenuifolium*	2601	2917	3174	3146	3038	3023	3085	3222
16	*Protium panamense*	2743	2911	3158	3138	3033	2984	3121	3219
17	*Swartzia simplex*	2711	2882	2991	3067	3091	3144	3183	3212
18	*Beilschmiedia pendula*	2375	2776	2962	2893	2616	2372	2248	2119
19	*Poulsenia armata*	3422	2766	2149	1792	1405	1165	996	859
20	*Rinorea sylvatica*	2577	2685	2743	2668	2524	2535	2544	2551

Let us focus on those species that have experienced severe population crashes, i.e., decreases of their abundance greater than 50% with respect to the first census in 1982. They comprise three species: *Psychotria horizontalis*, *Piper cordulatum* and *Poulsenia armata* (see figure 6.3(a)). *Piper cordulatum* is the species that experienced the most drastic bust; after an initial population growth from 1982 to 1985 its abundance crashed by almost 100% for the 5th census in 2000.

Figure 6.3(b) shows the variation of the species abundances versus the intraspecific interaction coefficient $M_{ii}(c)$ for the eight censuses. Notice that in general they parallel the evolution of the corresponding abundance trajectory $N_i(c)$ for $i = 1,2,\ldots,20$. In particular, the three crashing species also exhibit drastic drops of their effective intraspecific competition coefficients $M_{ii}(c)$.

For these three species, in order to make a standardized comparison between the drops of $N_i(c)$ and $M_{ii}(c)$, we computed the percentage variation along censuses 2 to 8 with respect to census 1 of the two quantities. Figure 6.4 shows that the drop of $M_{ii}(c)$ is steeper than the corresponding one for the $N_i(c)$. Notice that for all the three species the earliness of a 50% or larger decrease of $M_{ii}(c)$ with respect to the one for $N_i(c)$ is 15 years for *Psychotria horizontalis*, 5 years for *Piper cordulatum* and 10 years for *Poulsenia armata*.

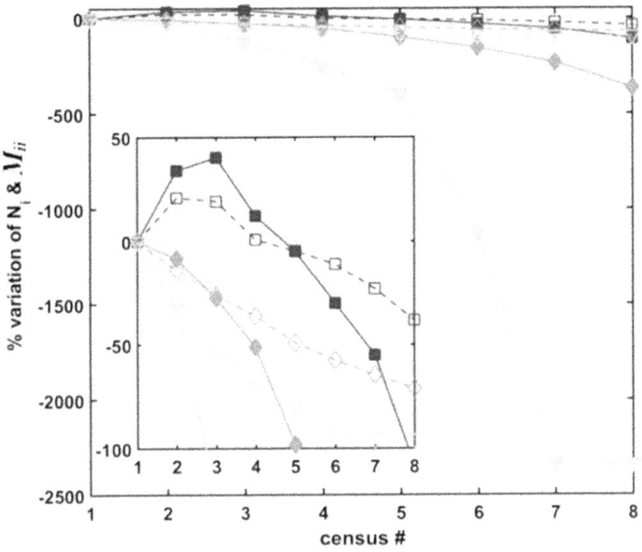

Figure 6.4. Percentage variations of the abundances (dashed lines) and intraspecific competition coefficients (filled lines) for the three species that experienced steep declines along the eight censuses, with respect to the first census (1982). Species # 7 *Psychotria horizontalis* (diamonds), species # 10 *Piper cordulatum* (triangles) and *Poulsenia armata* species # 19 (squares). The inset is a zoom-in plot to allow comparisons of the drops in both variables for each species. Reprinted from Fort and Grigera (2021b), copyright (2021), with permission from Elsevier.

Thus, the effective intraspecific interaction coefficient M_{ii}, estimated via MaxEnt, provides a clear early warning indicator of impending population crashes for species of trees that experimented population reductions of more than 50% in the number of individuals. They all exhibit a steeper earlier drop of $M_{ii}(c)$ than their abundances $N_i(c)$ (at least five years before).

As we have shown in section 6.4, the fact that the intraspecific elements of M_{ij} are much larger than the interspecific elements is because same happens for the covariance matrix, i.e., the variances σ_i^2 dominate over the covariances. We mentioned that the spatial variance has been used as an early warning signal in ecology. Figure 6.5 compares for the three crashing species the evolution of $M_{ii}(c)$ and that of $\sigma_i^2(c)$. Notice that the drop of $M_{ii}(c)$ is steeper than the corresponding one for $\sigma_i^2(c)$.

In fact, the factor $\dfrac{1 + (S - 2)\varepsilon}{1 + (S - 2)\varepsilon - (S - 1)\varepsilon^2}$ appearing in equation (6.4) for ε varying empirically among censuses between 0.046 and 0.058 (the values listed in table 6.1), is always $\approx$1. Therefore, the average (over species) intraspecific effective competition coefficient $\langle M_{ii} \rangle$ is well approximated by $-\dfrac{1}{\langle \sigma_i^2 \rangle}$. This suggests that the intraspecific effective competition coefficients can be approximated by the inverse of the variances. Table 6.3 shows the percentage difference of $1/\sigma_i^2$ relative to M_{ii} for the 20 most abundant species across the eight censuses. Notice that this percentage

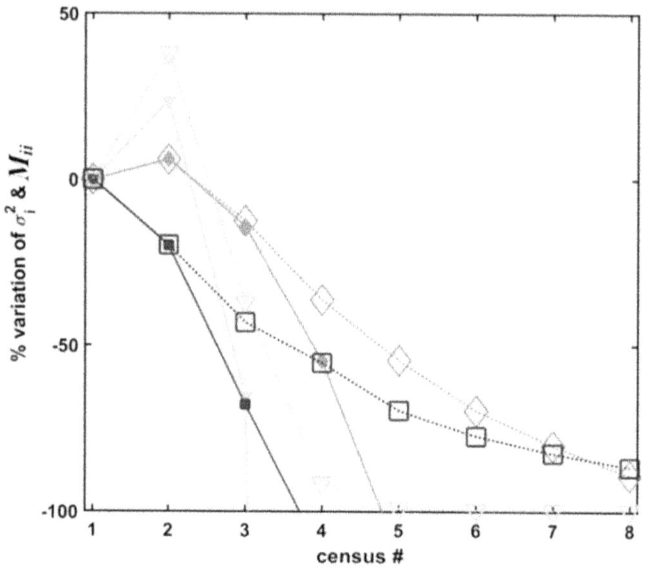

Figure 6.5. Percentage variations of the variances (dotted lines) and intraspecific competition coefficients (filled lines) for the three species that experienced steep declines along the eight censuses, with respect to the first census (1982). Species # 7 *Psychotria horizontalis* (diamonds), species # 10 *Piper cordulatum* (triangles) and *Poulsenia armata* species # 19 (squares). Reprinted from Fort and Grigera (2021b), copyright (2021), with permission from Elsevier.

Table 6.3. Percentage of difference of $1/\sigma_i^2$ relative to $M_{ii}$and Taylor's exponent in equation (6.9) for the 20 most abundant species (dbh > 10 mm). Highlighted in bold are those species which experimented population crashes.

Species name	Census (year)								Exponent ***b***
	1982	1985	1990	1995	2000	2005	2010	2015	
Hybanthus prunifolius	21	19	14	13	11	13	12	12	1.21
Faramea occidentalis	39	37	38	36	31	30	29	30	1.69
Trichilia tuberculata	22	23	23	23	24	24	27	24	1.29
Desmopsis panamensis	11	11	11	10	10	11	9	10	3.07
Alseis blackiana	15	15	13	12	11	10	9	9	0.52
Mouriri myrtilloides	34	36	33	30	25	24	25	30	1.72
Psychotria horizontalis	**18**	**17**	**18**	**16**	**14**	**13**	**12**	**13**	**1.78**
Hirtella triandra	31	28	26	24	22	23	24	24	1.81
Garcinia intermedia	11	11	12	10	9	10	11	11	0.66
Piper cordulatum	**16**	**22**	**21**	**11**	**5**	**4**	**5**	**5**	**1.46**
Capparis frondosa	20	20	18	15	14	12	13	13	1.69
Tetragastris panamensis	27	26	27	26	24	23	22	22	1.04
Sorocea affinis	15	15	14	14	13	14	16	14	1.33
Tachigali versicolor	19	19	21	18	15	12	6	7	1.37
Protium tenuifolium	34	37	35	34	35	39	42	44	1.27
Protium panamense	36	37	35	31	29	24	21	23	1.79

Swartzia simplex	15	15	16	15	14	15	15	15	1.39
Beilschmiedia pendula	28	22	23	24	24	25	23	22	2.72
Poulsenia armata	**46**	**40**	**39**	**39**	**34**	**33**	**32**	**29**	**1.47**
Rinorea sylvatica	5	5	4	4	4	4	4	4	2.28
Mean	23	23	22	20	19	18	18	18	

difference varies from all the species across the eight censuses between 4% and 46% (mean of 20%).

Hence, the early warning signal we are proposing is connected with the jump of the corresponding species variance (figure 6.5). This is an interesting result since an anomalous (large) variance is known to be a red flag anticipating an upcoming population drop (Fernández and Fort 2009, Donangelo *et al* 2010, Dakos *et al* 2010). That is, the variances may be used as early warnings in communities of coexisting species in cases in which covariances are not known. However, M_{ii}, which comes from the inverse of the covariance matrix, thus includes information from the whole community, i.e., more information than σ_i^2. And, as shown in figure 6.5, $M_{ii}(c)$ provides an earlier warning signal than $\sigma_i^2(c)$ of ongoing species population crashes.

At this point it is worth mentioning the well-known empirical Taylor's law (Taylor 1961), which states that the variance of species population density scales as a power-law function of the mean population density. In fact, all the 20 most abundant species verify that the variance for census # c scales with the population density for this census as

$$\sigma_i^2(c) = a_i \overline{x_i}(c)^{b_i}, \tag{6.9}$$

where $a_i > 0$ and the exponents b_i, obtained by linear regression, are listed in the last column of table 6.3.

Note that the b exponent for the three species which experience population busts are larger than 1 (see the last column of table 6.3). Thus, if we approximate the intraspecific competition coefficients by the inverse of the variances, from equation (6.9) we can express them as:

$$M_{ii}(c) \approx \frac{1}{\sigma_i^2(c)} = \frac{1}{a_i}\overline{x_i}(c)^{-b_i}. \tag{6.10}$$

From expression (6.10) it follows that

$$\frac{dM_{ii}(c)}{M_{ii}(c)} \approx -b_i \frac{d\overline{x_i}(c)}{\overline{x_i}(c)}, \tag{6.11}$$

and thus, since $b_i > 1$ for these three species (in fact $b_i > 1$ for 18 out of the 20 considered species), the relative changes of the intraspecific effective competition coefficients are larger than the relative changes experimented by the population densities. This result confirms what we already know, namely that the drop of the intraspecific competition coefficient is steeper than the corresponding one for the abundance. It also offers an observable to monitor to anticipate impending population busts, i.e.,

$$\boxed{\frac{M_{ii}(c+1) - M_{ii}(c)}{M_{ii}(c)}.} \tag{6.12}$$

6.7 Conclusion, caveats and future developments

The MaxEnt principle can be viewed as an inference method in which we have to select some set of empirical observations as constraints. In this study we took the spatial covariance matrix of species' abundances in quadrats. The rationale for this was that pairwise species interactions are enough to capture the behavior of a tree community. In order to take this description seriously, the goodness of this choice must be tested. Indeed, it could be that the tangle of complex biotic and abiotic interactions cannot be correctly described by pairwise maximum entropy modelling. We have previously shown (Fort 2020, Fort and Grigera 2021a, 2021b) that these models provide a reasonable description of the BCI tree community. Furthermore, in a first approximation, the effect of the interspecific interactions can be neglected compared to the intraspecific ones.

We can summarize the main findings of this chapter as follows:

I. For each species i the absolute value of the intraspecific competition coefficients for all census c, $M_{ii}(c)<0$, is much larger than those of the interspecific interaction coefficients $M_{ij}(c)$ with $i\neq j$ (which are of both signs).

II. The above result, $|M_{ii}|\gg|M_{ij}|$, can be derived from a similar empirical relationship observed for the covariance matrices, i.e., $\Sigma_{ii}\equiv \sigma_i^2\gg|\Sigma_{ij}|$.

III. For those tree species that crashed or suffered steep population declines across the eight tree censuses the drop of M_{ii} is always steeper than the drop of the corresponding abundance N_i.

IV. By (II), the drop of M_{ii} is linked to the anomalous variance, which is a known early warning of incoming catastrophic shifts. However, the sudden increase in the intraspecific competition, besides being more illuminating from a conceptual population dynamics modelling perspective, it provides a sharper signal than the jump in σ_i^2.

Therefore, we conclude that monitoring the quantity of equation (6.12) for tree species across censuses serves as an early warning indicator of an impending population bust; steep declines in this observable always occur several years in advance of comparable population drops.

It turns out that both this early warning signal and the forecasting model for tree trajectories presented in section 6.5 (Fort and Grigera 2021a) points to two species which are very likely to experience further important population losses. These species are *Psychotria horizontalis* and *Poulsenia armata*. Unfortunately the 2020 census was canceled because of COVID-19; we will have to wait until the next census takes place to confirm or reject this prediction.

Let us conclude with an important limitation of this approach: that it warns us of the impending species busts but it doesn't tell us what we can do in order to prevent or reverse the ongoing loss of individuals of this species. To advance in remedial situations it is necessary to link the drop in the observable with the biological or physical processes affecting the particular species. More generally, this requires the development of models based on understanding the (dynamical) processes that result in a system behaving the way it does.

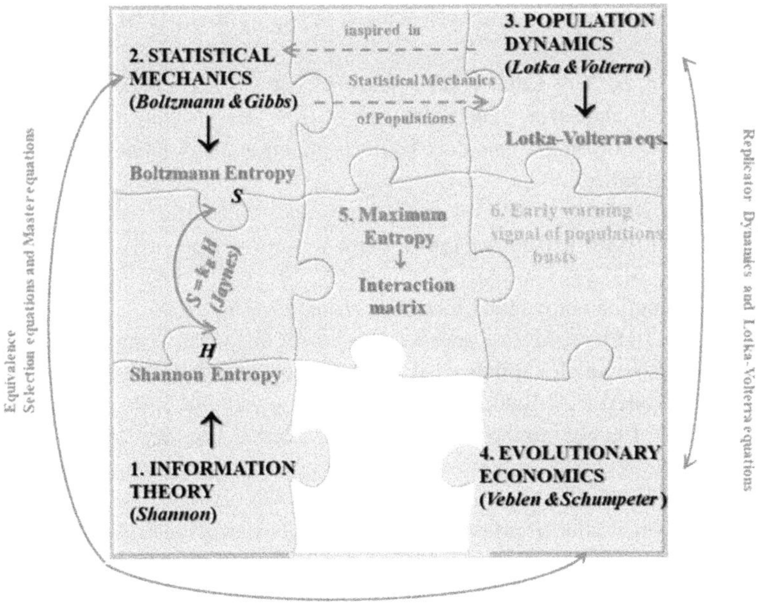

References

Adler P B, HilleRisLambers J and Levine J M 2007 A niche for neutrality *Ecol. Lett.* **10** 95–104

Adler P B *et al* 2018 Competition and coexistence in plant communities: intraspecific competition is stronger than interspecific competition *Ecol. Lett.* **21** 1319–29

Boettiger C, Ross N and Hastings A 2013 Early warning signals: the charted and uncharted territories *Theor. Ecol.* **6** 255–64

Carpenter S R and Brock W A 2006 Rising variance: a leading indicator of ecological transition *Ecol. Lett.* **9** 311–18

Carpenter S R *et al* 2011 Early warnings of regime shifts: a whole-ecosystem experiment *Science* **332** 1079–82

Chase J M 2014 Spatial scale resolves the niche versus neutral theory debate *J. Veg. Sci.* **25** 319–22

Chesson P 2000 Mechanisms of maintenance of species diversity *Annu. Rev. Ecol. Syst.* **31** 343–66

Clark J S and McLachlan J S 2003 Stability of forest biodiversity *Nature* **423** 635–8

Condit R 1998 Species distribution maps from the 50 ha plot at Barro Colorado Island *Tropical Forest Census Plots. Environmental Intelligence Unit* (Berlin: Springer)

Condit R *et al* 2017 Demographic trends and climate over 35 years in the Barro Colorado 50 ha plot *Forest Ecosyst.* **4** 17

Condit R, Perez R, Aguilar S, Lao S, Foster R and Hubbell S P 2019 Complete data from the Barro Colorado 50-ha plot: 423617 trees, 35 years, Dryad, Dataset, 2019 version

Dakos V, van Nes E H, Donangelo R, Fort H and Scheffer M 2010 Spatial correlation as leading indicator of catastrophic shifts *Theor. Ecol.* **3** 163–74

Donangelo R, Fort H, Dakos V, Scheffer M and van Nes E H 2010 Early warnings for catastrophic shifts in ecosystems: comparison between spatial and temporal indicators *Int. J. Bifurc. Chaos* **20** 315–21

Evans M R 2012 Modelling ecological systems in a changing world *Phil. Trans. R. Soc.* **B367** 181–90

Fernández A and Fort H 2009 Catastrophic phase transitions and early warnings in a spatial ecological model *J. Stat. Mech.* **2009** P09014

Fort H 2020 *Ecological Modelling and Ecophysics: Agricultural and Environmental Applications* (Bristol: IOP Publishing)

Fort H and Grigera T 2021a A method for predicting species trajectories tested with trees in Barro Colorado tropical forest *Ecol. Model.* **446** 109504

Fort H and Grigera T 2021b A new early warning indicator of tree species crashes from effective intraspecific interactions in tropical forests *Ecol. Indic.* **125** 107506

Fort H and Segura A 2018 Competition across diverse taxa: quantitative integration of theory and empirical research using global indices of competition *Oikos* **127** 392–402

Garzon-Lopez C X, Jansen P A, Bohlman S A, Ordonez A and Olff H 2014 Effects of sampling scale on patterns of habitat association in tropical trees *J. Veg. Sci.* **25** 349–62

Hooper D U *et al* 2012 A global synthesis reveals biodiversity loss as a major driver of ecosystem change *Nature* **486** 105–8

Hubbell S P 2008 Approaching tropical forest complexity, and ecological complexity in general, from the perspective of symmetric neutral theory *Tropical Forest Community Ecology* ed W Carson and S Schnitzer (New York: Wiley)

Hubbell S P 2009 Neutral theory and the theory of island biogeography *The Theory of Island Biogeography at 40: Impacts and Prospects* ed J Losos and R E Ricklefs (Princeton, NJ: Princeton University Press)

Ishida A *et al* 2003 Leaf physiological adjustments to changing lights: partitioning the heterogeneous resources across tree species *Pasoh: Ecology of a Lowland Rain Forest in Southeast Asia* (Berlin: Springer)

Ives A R and Carpenter S R 2007 Stability and diversity of ecosystems *Science* **317** 58–62

Jaynes E T 1985 Bayesian methods: general background *Maximum Entropy and Bayesian Methods in Applied Statistics* ed J H Justice (Cambridge: Cambridge University Press) pp 1–25

MacDougall A S, McCann K S, Gellner G and Turkington R 2013 Diversity loss with persistent human disturbance increases vulnerability to ecosystem collapse *Nature* **494** 86–90

Millennium Ecosystem Assessment 2005 *Ecosystems and Human Well-being: Synthesis Report* (Washington, DC: Island Press)

Scheffer M and Carpenter S R 2003 Catastrophic regime shifts in ecosystems: linking theory to observation *Trends Ecol. Evol.* **12** 648–56

Shmueli G and Lichtendahl K C 2016 *Practical Time Series Forecasting with R: A Hands-on Guide* 2nd edn (Green Cove Springs, FL: Axelrod Schnall Publishers)

Taylor L R 1961 Aggregation, variance and the mean *Nature* **189** 732–35

Thompson R and Starzomski B M 2006 What does biodiversity actually do? A review for managers and policy makers *Biodivers. Conserv.* **16** 1359–78

Volkov I, Banavar J R, Hubbell S P and Maritan A 2009 Inferring species interactions in tropical forests *Proc. Natl Acad. Sci.* **106** 13854–9

Whittaker R H 1975 *Communities and Ecosystems* (New York: MacMillan)

Chapter 7

Modelling markets as ecosystems with the help of maximum entropy

'Investing should be dull, like watching paint dry or grass grow. If you want excitement, take $800 and go to Las Vegas.'

—Paul Samuelson

We start with a short history of different attempts of modelling the dynamics of markets and of finding a law for the price fluctuations. We briefly review the conceptual path from Louis Bachelier's theory of Brownian motion to the modern efficient-market hypothesis (EMH), which states that asset prices reflect all available information and, as a consequence, it is impossible to 'beat the market' consistently.

Next we discuss the criticism to the EMH. One alternative economic theory to EMH is the adaptive market hypothesis (AMH), which extends the main tenets of the controversial EMH by applying the principles of evolution and behavior to financial interactions.

Then we move onto the aim of this chapter which is, by adopting the AMH point of view, to develop a method of market dynamics forecasting that combines different previously discussed concepts of physics, biology and economics. This method is based on the replicator dynamics (RD) (discussed in chapter 4) as an equation to model natural selection in markets. A main difficulty is how to obtain the payoff matrix connecting the pairwise effects between interacting market entities. Thus we use the pairwise maximum-entropy (PME) procedure (introduced in chapter 5) to estimate the payoff matrix. The resulting method is called replicator dynamics pairwise maximum entropy (RDPME).

To test the method, daily market values from 2014 to 2019 of America's top revenue companies are used. As it is customary in time series forecasting analysis, these series are divided into a training period, used to infer the RDPME parameters (intrinsic growth rates and payoff matrix), and a validation period, used to validate the model. Different partitions into training and validation periods are considered.

We show that the RDPME method outperforms the stochastic benchmark of the geometric random walk in predicting empirical shares x_i for most of the companies along most choices of validation periods (although the mean relative errors of the RDPME are in general only modestly smaller than those produced by the stochastic benchmark).

We conclude this chapter by reviewing some caveats, extensions, and possible future improvements.

7.1 Background: a short history of market modelling

7.1.1 Of pollen motion, drunks and bond prices

The French stock broker Jules Augustin Frédéric Regnault was the first who suggested a modern theory of stock price changes, based on statistical and probabilistic analyses, in his *Calcul des Chances et Philosophie de la Bourse* (1863). His hypotheses were later used by the French mathematician Louis Bachelier to found the science of investing or mathematical finance with his 1900 doctoral thesis, *Théorie de la speculation* (The Theory of Speculation). It was an attempt to create a formula to capture the movement of bonds in the Paris *Bourse*, or stock exchange. Bachelier developed the theory of what we nowadays call ***Brownian motion***, named after the Scottish botanist Robert Brown.

In 1827, Brown was struck by the jittery motion of small particles of pollen floating on the surface of water he observed through a microscope. Since pollen was organic, Brown's first thought was that it might be exhibiting signs of life as it jumped around the surface. But when Brown later saw the pollen's behavior replicated by inorganic matter like coal dust, he ruled out the hypothesis that the effect was life-related. Brown couldn't provide an explanation of this frenetic dance.

In fact, it was Einstein in 1905 who developed the theory for the Brownian motion by modelling the motion of the pollen particles as being displaced by collisions with individual water molecules. The direction of the force of atomic bombardment is constantly changing, and at different times the particle is hit more on one side than another, leading to the seemingly random nature of the motion. Importantly, this explanation of Brownian motion served as convincing evidence that atoms and molecules exist. The many-body interactions that yield the Brownian motion cannot be solved by taking into account every involved molecule. Instead, Einstein adopted a statistical mechanics approach, in terms of a probabilistic description of the molecular effects on the grain of pollen. He arrived at a theory formally identical to the one devised five years earlier by Bachelier for the movement of bonds. So here we have another formal equivalence linking statistical mechanics with finance economics.

When treated as a discrete-space (integers) and discrete-time model, Brownian motion becomes the epitome of a stochastic or random process: the ***random walk***. In other words, essentially, Bachelier's formula tells us that the price of a security rises or falls accordingly to the result of flipping a coin. This random walk is also called the ***drunkard's walk***. The simplest example is one-dimensional; let us imagine a drunkard who exits a bar attempting to return home. He is so intoxicated that he only remembers his house and the bar are both on the same street, but he has no idea whether it is to the left or to the right of the bar. So he starts by arbitrarily choosing the right and giving a step toward this direction. But then he hesitates and before every next step he faces a binary decision: right or left? This concept was introduced into science by Karl Pearson in a letter to *Nature* in 1905: 'A man starts from a point O and walks l yards in a straight line; he then turns through any angle whatever and walks another l yards in a second straight line. He repeats this process n times.' Pearson then required the probability that after these n stretches the man is at a distance d from his starting point O. Interestingly, the solution to this problem was provided in the same volume of *Nature* by Lord Rayleigh (1905), who told him that he had solved this problem 25 years earlier when studying the superposition of sound waves of equal frequency and amplitude but with random phases. Pearson concluded that 'The lesson of Lord Rayleigh's solution is that in open country the most probable place to find a drunken man who is at all capable of keeping on his feet is somewhere near his starting point!'

However, Bachelier's theory remained unnoticed gathering dust on library bookshelves for 60 years until the American economist Paul Samuelson discovered it. Samuelson expanded and refined Bachelier's work. As Bachelier did, he also assumed that market prices are the best estimates of value and that price changes follow random patterns, so future news and stock prices are unpredictable. A problem with Bachelier's modelling was that security prices can take negative values, but this is nonsense from a financial point of view. This drawback was overcome by Samuelson (1965a, 1965b) who proposed the so-called *Geometric Brownian* motion, in which the *natural logarithm* of the price is assumed to walk a random walk, usually a random walk with drift. That is, the changes in the natural log from one period to the next are assumed to be independent and identically normally distributed.

Therefore, the so-called *random walk hypothesis* (RWH) (Cootner 1964, Fama 1965, Malkiel 1973) is a financial theory stating that stock market prices evolve according to a random walk (so price changes are random) and thus cannot be predicted.

7.1.2 The efficient market hypothesis and the crypto-trading hamster Mr Goxx

Since prices move as a random walk, Bachelier set forth the revolutionary conclusion that there is no useful information contained in historical price movements of securities. In his own words:

'The mathematical expectation of a speculator is nil' (Bachelier 1900).

Between the time Bachelier proposed his theory and the 1960s when it was rediscovered, the Wall Street Crash of 1929 occurred and a severe worldwide economic depression, the so-called Great Depression, took place during the 1930s. These remarkable economic events stimulated some fruitful investigation. For example, James Case in his 2008 *Competition: The Birth of a New Science* tells the findings of the American economist and businessman Alfred Cowles III supporting the difficulties of predicting market movements:

'In 1928, he purchased subscriptions to twenty-four of the most widely circulated newsletters and monitored their performance for four eventful years–... To his surprise, Cowles discovered that exactly none of twenty-four publications had managed to foretell either the crash of 1929 or the steady market decline that followed'

This idea that financial market returns are difficult to predict is the basis of the modern *efficient-market hypothesis* (EMH) (Samuelson 1965a, 1965b, Fama 1970), which states that asset prices reflect all available information and as a consequence that it is impossible to 'beat the market' consistently since market prices should only react to new information. In other words, stocks always trade at their fair value on exchanges, making it impossible for investors to purchase undervalued stocks or sell stocks for inflated prices. Therefore, it should be impossible to outperform the overall market through expert stock selection or market timing. The RWH is consistent with the EMH.

EMH is somehow related with the rational choice theory we discussed in chapter 4, both assume that there is a well-defined model of economic behavior and that rational investors would all follow it. EMH added another step. In the strong version of the theory, financial markets, because they are populated by a multitude of rational and competitive players, would always set prices that reflected all available information in the most accurate possible way.

Proponents of EMH posit that investors benefit from investing in a low-cost, passive portfolio. In fact, data compiled by Morningstar Inc., in its August 2019 Active/Passive Barometer study, supports the EMH. This report measures the performance of U.S. active funds against passive peers in their respective Morningstar Categories. The Active/Passive Barometer spans more than 4000 unique funds that account for approximately $12.5 trillion in assets, or about 64% of the U.S. fund market. Morningstar compared active managers' returns in all categories against a composite made of related index funds and exchange-traded funds (ETFs). The study found that over a 10 year period beginning June 2009, only 23% of active managers were able to outperform their passive rivals (Morningstar 2019).

In addition, while a percentage of active managers do outperform passive funds at some point, the challenge for investors is being able to identify which ones will do so over the long term. Less than 25% of the top-performing active managers can consistently outperform their passive manager counterparts over time (Downey 2021).

Economist Burton Malkiel mocked the financial services industry. He famously quipped, 'A blindfolded monkey throwing darts at a newspaper's financial pages could select a portfolio that would do just as well as one carefully selected by the experts.' Actually, not a monkey but a hamster is currently proving such a claim. And it's not stocks: it's cryptocurrencies. This crypto-trading hamster that goes by the name of *Mr Goxx* is a social media sensation in Germany. Named after Mt Gox—a Tokyo-based cryptocurrency exchange company founded in 2010 that was responsible for more than 70% of bitcoin transactions at its peak until it was hacked and declared bankruptcy in 2014—the rodent is equipped with a trading office attached to his regular cage. Mr Goxx goes on the wheel where he selects which cryptocurrency he would like to trade by simply spinning it. The office has two tunnels: one for buying and one for selling, and every time he goes through one of them, the desired transaction is completed. The hamster livestreams his crypto tradings on video live streaming service *twitch*. As of

24 September 2021, the hamster's performance was up more than 16%[1] since it began trading in June 2021 (Protos 2021), an impressive feat for anyone. Bitcoin and the S&P 500 have climbed about 14% and 5% respectively, over the same time period. In other words, the hamster is beating not only bitcoin, but the S&P 500, as well as many professional traders and funds since it started trading.

7.1.3 Criticism to the efficient market hypothesis

There have been several criticisms of EMH. Some of the most common are as follows.

First, real markets exhibit inefficiencies and there are some markets that are less efficient than others. An inefficient market is one in which an asset prices do not accurately reflect their true or *fair value*. This may occur for several reasons. Market inefficiencies may exist due to information asymmetries, lack of competition, tax distortions, lack of buyers and sellers (i.e., low liquidity), high transaction costs or delays, market psychology, and human emotion, among other reasons (Downey 2021). In particular, psychology and emotions are key drivers of *market sentiment*. That is, the overall consensus of investors about a stock or the stock market as a whole is not always based on fundamentals. Day traders and technical analysts also rely on market sentiment, as it influences the technical indicators they utilize to measure and profit from short-term price movements often caused by investor attitudes toward a security. In *A Study of History*, Arnold Toynbee (1946) warns to beware of what he calls the 'apathetic fallacy', whereas 'Ruskin warned his readers against the 'pathetic fallacy' of imaginatively endowing inanimate objects with life', when addressing a system composed of human beings we need to avoid the converse error of blindly applying a scientific method devised for the study of inanimate nature.

Second, Eugene Fama never imagined that his efficient market would be 100% efficient all the time. That would be impossible, as it takes time for stock prices to respond to new information. The EMH, therefore doesn't give a strict definition of how much time prices need to revert to fair value. Third, the EMH assumes a sort of homogeneity of the market participants; i.e., all investors perceive all available information in precisely the same manner. But this is hardly the case. For example, there are many different methods for analyzing and valuing stocks. If one investor focuses on value, and then looks for undervalued market opportunities, while another focuses on growth and evaluates the same stock on the basis of its growth potential, these two investors may arrive at a different assessment of the stock's fair market value. Therefore, since investors value stocks differently, it is impossible to determine what a stock should be worth under an efficient market. Fourth, under the EMH, no investor should ever be able to beat the market or the average annual returns that all investors and funds are able to achieve using their best efforts. This is why proponents of the EMH conclude investors may profit from investing in a low-cost, passive portfolio, i.e., purchasing a representative benchmark, such as the S&P

[1] https://twitter.com/mrgoxx/status/1441999068046233602?s=20.

500 index, and hold it over a long time horizon. But there are many investors who have consistently beaten the market. Warren Buffett is one of those who's managed to outpace the averages year after year (Dhir 2021).

In addition, there is evidence against the RWH for the logarithm of wealth relatives of small-firms portfolios (Lo *et al* 2002).

7.1.4 Combining efficient market hypothesis with behavioral finance: the adaptive market hypothesis

Opponents of EMH believe that it is possible to beat the market and that stocks can deviate from their fair market values. In particular, behavioral finance challenges the notion that companies always trade at their fair value, pointing out that investors are not always rational and stocks do not always trade at their fair value during financial bubbles, crashes, and crises. Thus behavioral economists attempt to explain stock market anomalies through psychology-based theories. This is the case of the adaptive market hypothesis (AMH), an alternative economic theory that combines principles of the well-known and often controversial EMH with behavioral finance (Lo 2004).

AMH argues that people are motivated by their own self-interest, make mistakes, and tend to adapt and learn from them. In fact, Andrew Lo, the theory's founder, believes that people are mainly rational, but sometimes can overreact during periods of heightened market volatility (Liberto 2021). He postulates that investor behaviors—such as loss aversion, overconfidence, and overreaction—are consistent with evolutionary models of human behavior, which include actions such as competition, cooperation, adaptation, and natural selection (Lo 2004). In the next subsection we will briefly elaborate on this apparent dichotomy of competition/cooperation in markets.

7.1.5 Competition (and cooperation) in financial markets

Competition has been for a long time the keyword of financial markets. There are several kinds of competition. There is competition between investment funds for money, or between individual investors, and between firms to sell their products. Indeed, for several years strategic management researchers focused on competitive relationships between firms (Porter 1980).

However, businesses also cooperate with each other, in various ways and forms. Since the early 1980s, we have been able to observe a soaring interest in alliances, joint ventures, collusions, federations, clusters and so on (Czakon 2010), collectively called interorganizational relationships (for a review see: Oliver and Ebers 1998). Actually, cooperation and competition characterize the inter-firm relationships in strategic alliances (Clarke-Hill *et al* 2003). Therefore, the competitive paradigm (Rumelt 1991) as well as the cooperative paradigm (Dyer and Singh 1998) both offer incomplete pictures of a more complex tangle of relationships between firms. Instead, the neologism '*Coopetition*' (Yami *et al* 2010), coined to describe cooperative competition, seems to better capture the reality of inter-firm relationships.

Inter-firm dynamics can be classified as competitive, cooperative, mixed (competitive-cooperative), or neither/nor (Chen 2002). A competitive action may elicit a cooperative response (from players in either the same industry or a different industry from where the action is taken), and cooperation between two firms often provokes competitive actions. As we have seen in chapter 3, this mix of competition and cooperation is quite common in ecological communities thus suggesting that a more realistic picture of markets is provided by networks ecosystems (Astley and Fombrun 1983, Emary and Fort 2021, Oliver and Ebers 1998).

Interestingly, coopetition is a dynamic process; a firm's competition with another firm at one level, at a given time and within one organizational unit, usually may turn cooperation with this same firm in different circumstances and vice versa (Chen 2002).

Often coopetition takes place when companies that are in the same market work together in the exploration of knowledge and research of new products, at the same time that they compete for market share of their products and in the exploitation of the knowledge created. In this case, the interactions occur simultaneously and in different levels in the value chain. For instance, on 17 March 2020, Pfizer Inc. and BioNTech SE announced a collaboration to jointly develop a COVID-19 vaccine (Pfizer 2020). This is an example of coopetition agreement between the two companies that increased the manufacturing capacity to meet the global supply for the vaccine. In this way the companies were able to produce millions of vaccine doses by the end of 2020 and hundreds of millions of additional doses in 2021. later on, in June 2021, the companies announced another deal with the government to provide 500 million more doses of the vaccine to support some of the poorest countries. The agreement states that 60% of the doses are to be purchased during the first half of 2022 (Pfizer 2021). BioNTech contributed the vaccine candidates, while Pfizer contributed the clinical research and development as well as the manufacturing and distribution capabilities of the company.

At any rate, whatever the pattern of inter-firm relationships, at the end of the day money tends to flow from firms that underperform to those that outperform the market. This suggests that the concepts of population dynamics and organizational ecology we discussed in chapter 4 can be useful to describe market dynamics (Dosi and Nelson 1994, Moore 1993a, 1993b, Nelson and Winter 1982), which is our goal, as we explain in the next section. For example, Lakka *et al* (2013) analyze the evolutionary and competitive dynamics of the highly concentrated desktop/laptop operating systems market as if companies were species of an ecosystem and draws conclusions that may be valuable inputs for managerial decisions and strategic planning to the players of software markets.

7.2 Goal

Our goal is, adopting the AMH point of view, to develop a method of market dynamics forecasting that combines different previously discussed concepts of physics, biology and economics.

As we have seen in chapter 4, natural selection is concerned with the differential rates of expansion ('fitness') of the competing, interacting members of a population. And, according to evolutionary economics, a parallelism between phenotypes and companies in a market can be traced. The economic analogue of natural selection among firms can be thought as operating as the market determines which firms are profitable and which are unprofitable, and tends to push out of business the latter. The notion of fitness of a firm, reflecting its relative competitiveness, is what determines its chances of growth and survival. Therefore, as a result of selection, the companies with greatest fitness capture market share.

It thus seems natural to regard markets dynamics as an example of frequency dependent selection and then model their evolution through the RD. In fact, even though many papers using RD have been published to model market evolution (see for instance Mazzucato 2000 and references therein), Cantner (2017) noticed that it is quite astonishing that empirical attempts trying to answer the question of whether market selection is operating as proposed by evolutionary theory are rare. Furthermore, as far as we know, no estimations of the whole payoff matrix of the RD involving several companies from empirical data have been carried out yet. Thus, let us develop a general procedure to quantitatively estimate the goodness of the RD for modelling market dynamics.

7.3 Data

Based on the above ideas we build here an evolutionary model for firms of the New York Stock Exchange (NYSE), the world's largest stock exchange by market capitalization of its listed companies. In fact, since there are 2800 listed firms in NYSE, working with all of them would be a daunting task. Thus, here we work with a set of main firms by their revenues. Specifically, the firms were selected such that:

- They are all amongst companies with the largest revenues in the Fortune 500 list as of 29 March 2018 (Fortune 2018).
- They simultaneously were in the list for six years in a row, from 2014 to 2019[2].
- The market values[3] of these firms were available in the 2014–2019 Fortune 500 lists[4].

[2] Thus, for example a firm like Dell Technologies, ranked 35 in the 2018 Fortune 500 list (Fortune 2018), wasn't included because from 2014 to 2016 it wasn't in the Fortune 500 list.

[3] Throughout this chapter we will use indistinctly 'market value' and 'market capitalization' or 'market cap' for short. Market capitalization is basically the number of a company's shares outstanding multiplied by the current price of a single share. More rigorously speaking, market value is more amorphous and more complicated, assessed using numerous metrics and multiples, such as price-to-earnings, price-to-sales, and return-on-equity (Boyte-White 2020).

[4] For example, there is no market value listed for State Farm Insurance in 2018 (Fortune 2018) or 2019 (Fortune 2019).

In this way we obtained the set of 38 firms of table 7.1. The market value of this set of companies summed \$8.05 trillion as of 29 March 2018 (Fortune 2018, Quandl 2019) representing 27.5% of the 29.6 trillion market value of all the listed firms at NYSE (NYSE 2021). Furthermore, over the six years it represented at least 25% of

Table 7.1. Companies considered in this study ordered by their market value as of 11 October 2019.

Firm #	Name	Ticker	Market Value v in \$M 10/11/2019
1	Apple	AAPL	1067,476.0
2	Microsoft	MSFT	1066,514.0
3	Amazon	AMZN	856,704.6
4	Alphabet	GOOGL	842,971.3
5	Berkshire Hathaway	BRK	510,842.2
6	JP Morgan	JPM	371,355.9
7	Johnson & Johnson	JNJ	346,601.6
8	Walmart	WMT	341,996.7
9	P&G	PG	302,998.6
10	Exxon Mobil	XOM	291,861.7
11	AT&T	T	274,597.1
12	Bank of America	BAC	269,103.0
13	Home Depot	HD	256,988.6
14	Verizon	VZ	247,856.4
15	Chevron	CVX	220,501.3
16	Wells Fargo	WFC	216,824.5
17	Boeing	BA	210,971.2
18	UnitedHealth Group	UNH	210,451.4
19	Comcast	CMCSA	207,024.1
20	Citigroup	C	158,359.9
21	Costco	COST	130,841.9
22	IBM	IBM	126,467.5
23	Lowe’s	LOW	85,523.0
24	CVS Caremark	CVS	81,853.3
25	General electrics	GE	76,798.2
26	Anthem	ANTM	60,769.4
27	Target	TGT	57,137.1
28	General motors	GM	50,784.3
29	Walgreens Boots Alliance	WBA	48,760.7
30	Phillips 66	PSX	47,783.1
31	Marathon Petroleum	MPC	41,750.6
32	Valero Energy	VLO	36,467.5
33	Ford Motors	F	35,030.8
34	McKesson	MCK	24,914.0
35	Kroger	KR	19,439.3
36	AmerisourceBergen	ABC	17,074.4
37	Cardinal Health	CAH	13,988.4
38	Fannie Mae+Freddy Mac	FNMA+FMCC	6336.7

the total NYSE market cap in the period 2015–2019 (NYSE 2021). Therefore, this set constitutes a representative sample of the market as well as being much more manageable.

The dataset consists of time series of the market values or these 38 spanning 1455 days, from 2 January 2014 to 11 October 2019 (Quandl 2019), i.e., $v_i(t)$, with $i = 1,2, \ldots,38$ and $t = 1,2,\ldots,1455$.

Since it seems natural to assume that the fitness is frequency dependent, we will assume that the Replicator Dynamics equation (RDE) describes the dynamics of market shares. Here we equate market shares with market value frequencies, given for time t, as:

$$x_i(t) = \frac{v_i(t)}{\sum_{j=1}^{38} v_j(t)} \quad i = 1,2,\ldots,38, \tag{7.1}$$

which, by construction, verify for all time t that:

$$\sum_{i=1}^{38} x_i(t) = 1. \tag{7.2}$$

7.4 Modelling: replicator dynamics combined with pairwise maximum entropy or RDPME model

7.4.1 Frequency dependent evolutionary model

To model the dynamics of the chosen set of firms, representative of the NYSE market, we use a finite difference version of the RDE equation (4.10), with time steps coinciding with days, given by:

$$x_i(t+1) - x_i(t) = x_i(t)\left(\sum_{j=1}^{38} P_{ij}x_j - \phi(\mathbf{x};\, t)\right), \quad i = 1,2,\ldots,38, \tag{7.3}$$

where $\mathbf{P} = [P_{ij}]$ is a 38×38 *payoff matrix*, that is, P_{ij} is the payoff received by company i from its interaction with company j. So our first task is to estimate this matrix. To do this we will use once again the recipe developed in chapter 5, i.e., to infer through the MaxEnt principle a PME model. The interaction matrix $\mathbf{I} = [I_{ij}]$, given by equation (5.10), in turn yields a payoff matrix given by equation (4.23). Thus we finally arrive at the set of equations for the RDPME method:

$$x_i(t+1) - x_i(t) = r_i x_i(t)\left(\sum_{j=1}^{38} I_{ij}x_j - \phi(\mathbf{x};\, t)/r_i\right), \quad i = 1,2,\ldots,38, \tag{7.4}$$

7.4.2 Parameter estimation

From covariance matrices to interaction matrices

As is customary in time series forecasting analysis, we partition the data set of 38 time series of length T into an earlier training period T_{tr} and a later validation period T_v (such that $T_{tr} + T_v = T$) (Shmueli and Lichtendahl 2016). The training period is used to estimate the parameters of the model (in our case, the RDE), and then this model with these estimated parameters is used to generate forecasts to be compared with data corresponding to the validation period. In this way we assess the predictive performance of a model on new data. Varying the training period T_{tr} allows one to develop multiple model instances. Thus, the validation partition is used to assess the performance of each of these models for different time frames.

Specifically, we obtain the covariance matrix $\boldsymbol{\Sigma} = [\Sigma_{ij}]$ whose entries are given by (see equation (5.5b)):

$$\Sigma_{ij} = \sum_{t=1}^{T_{tr}} v_i(t)v_j(t)/T_{tr} - \sum_{t=1}^{T_{tr}} v_i(t) \sum_{s=1}^{T_{tr}} v_j(s)/T_{tr}^2, \tag{7.5}$$

and then, from equation (7.5), we first obtain matrix $\boldsymbol{M} = \boldsymbol{\Sigma}^{-1}$ through equation (5.9b) and next matrix $\mathbf{I}$ through equation (5.10).

Some general properties of matrix $\mathbf{I}$ are:

I. It has positive as well as negative elements; this is in agreement with what we mentioned in section 7.5.1.

II. Furthermore, positive elements dominate over negative elements. Indeed, the same happens for the covariance matrix, which is a known result (Chan *et al* 1999, Engle 2009, Vasicek and McQuown 1972).

III. Despite most of its entries being in (−0.5,+0.5), there is a large dispersion, as shown in panel (b) of figure 7.1 which shows this matrix $\mathbf{I}$ for a training period $T_{tr} = 1000$.

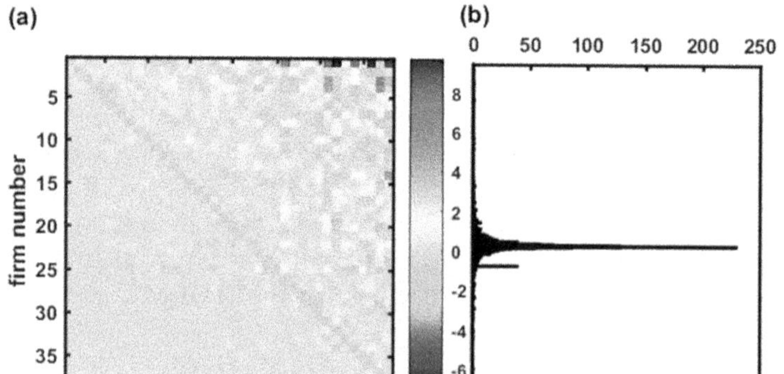

Figure 7.1. (a) Matrix $\mathbf{I}$ for T_{tr} =1000. (b) The corresponding histogram of its elements I_{ij} (the secondary peak at −1 corresponds to the diagonal elements).

IV. As can be seen from panel (a) of figure 7.1, all the large departures to these values fall in the upper triangle part; that is the interaction strength of smaller cap firms (high firm numbers) over larger firms (low firm numbers).

Property IV can be understood from the covariance matrix with firms ranked by their market value as follows:

First, from equation (7.5), the larger entries of $\mathbf{\Sigma}$ occur at the upper left corner (panel (a) of figure 7.2). Second, since $\boldsymbol{M} = -\mathbf{\Sigma}^{-1}$, the larger entries of this MaxEnt matrix will concentrate at the bottom right corner (panel (b) of figure 7.2). This can be shown by using that $\mathbf{\Sigma}^{-1} = \text{adj}(\mathbf{\Sigma})/\det(\mathbf{\Sigma})$ (equation (vi) of Box 5.1). Thus, for firms with larger market values $\text{adj}(\mathbf{\Sigma}) \ll \det(\mathbf{\Sigma})$; conversely, for firms with smaller market values $\text{adj}(\mathbf{\Sigma}) \gg \det(\mathbf{\Sigma})$. And this explains the places where the smaller (larger) values of $\mathbf{M}$ do occur.

Second, to obtain the MaxEnt interaction matrix $\mathbf{I}$ the entries of each row i of $\mathbf{M}$ are divided by M_{ii}; since M_{ii} increases with i this implies that the stronger interaction coefficients will appear in the upper right corner (panel (c) of figure 7.2).

We will use the RDPME equation (7.4) to predict the market shares for $T_{\text{tr}} + 1 \leqslant t \leqslant T = T_{\text{tr}} + T_{\text{v}}$ with an interaction matrix $\mathbf{I}$ estimated for $t = T_{\text{tr}}$. It is worth remarking that this matrix depends of the training period. Figure 7.3 shows the matrices $\mathbf{I}$ obtained for different training periods; in the left column $T_{\text{tr}} = 100$, 300 and 500 days, while in the right column $T_{\text{tr}} =1000$, 1200 and 1400.

By simple visual inspection it seems that $\mathbf{I}$ changes faster from lower values of T_{tr}. In the appendix at the end of this chapter we propose a metric to quantify the variation rate of an arbitrary matrix $\mathbf{M}$, and that indicates that after $T_{\text{tr}} = 600$ the variation rate for matrix $\mathbf{I}$ stabilizes thus suggesting the entrance of the market into a kind of stationary state.

Growth rates

To compare predictions from the RDE against the observed $x_i(t)$ for $t > T_{\text{tr}}$, we still have to estimate the growth rate r_i parameter. There are different alternative procedures to obtain this estimate from equation (7.4). One direct way is by taking:

$$r_i = \max\left(0, \frac{x_i(T_{\text{tr}}) - x_i(1)}{\sum_{t=1}^{T_{\text{tr}}-1} x_i(t)\left(\sum_{j=1}^{38} P_i x_j(t) j - \phi(\mathbf{x}; t)\right)}\right) \tag{7.6}$$

The maximum in equation (7.6) is because the sign of $x_i(t+1) - x_i(t)$ in the replicator dynamics equation (7.3) must be the same as the one of $\left(\sum_{j=1}^{38} P_{ij}x_j - \phi(\mathbf{x}; t)\right)$, i.e., it must be determined by whether the fitness of firm i is above or below the weighted average fitness, and thus r_i must be $\geqslant 0$ for all i.

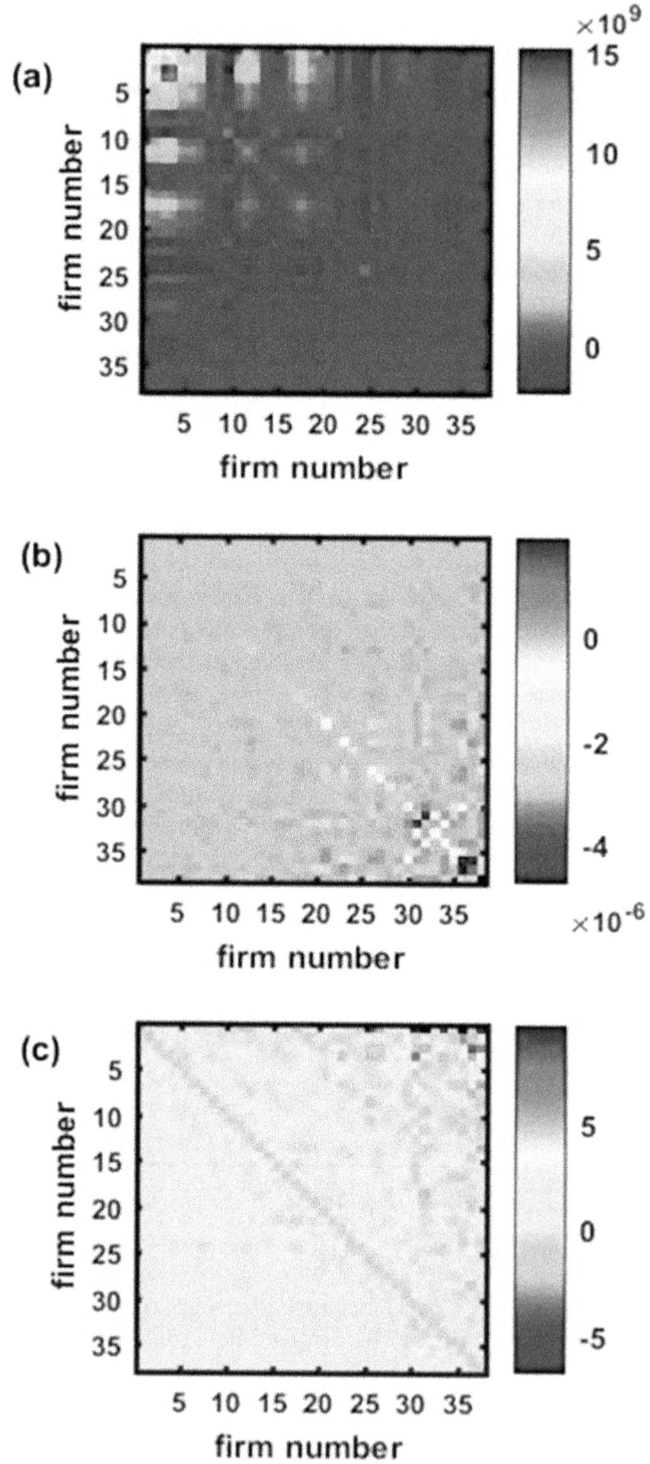

Figure 7.2. Covariance and MaxEnt interaction matrix for $T_{tr} = 1000$. (a) Covariance $\mathbf{\Sigma}$ matrix. (b) $\boldsymbol{M} = \mathbf{\Sigma}^{-1}$ matrix. (c) MaxEnt interaction matrix **I**.

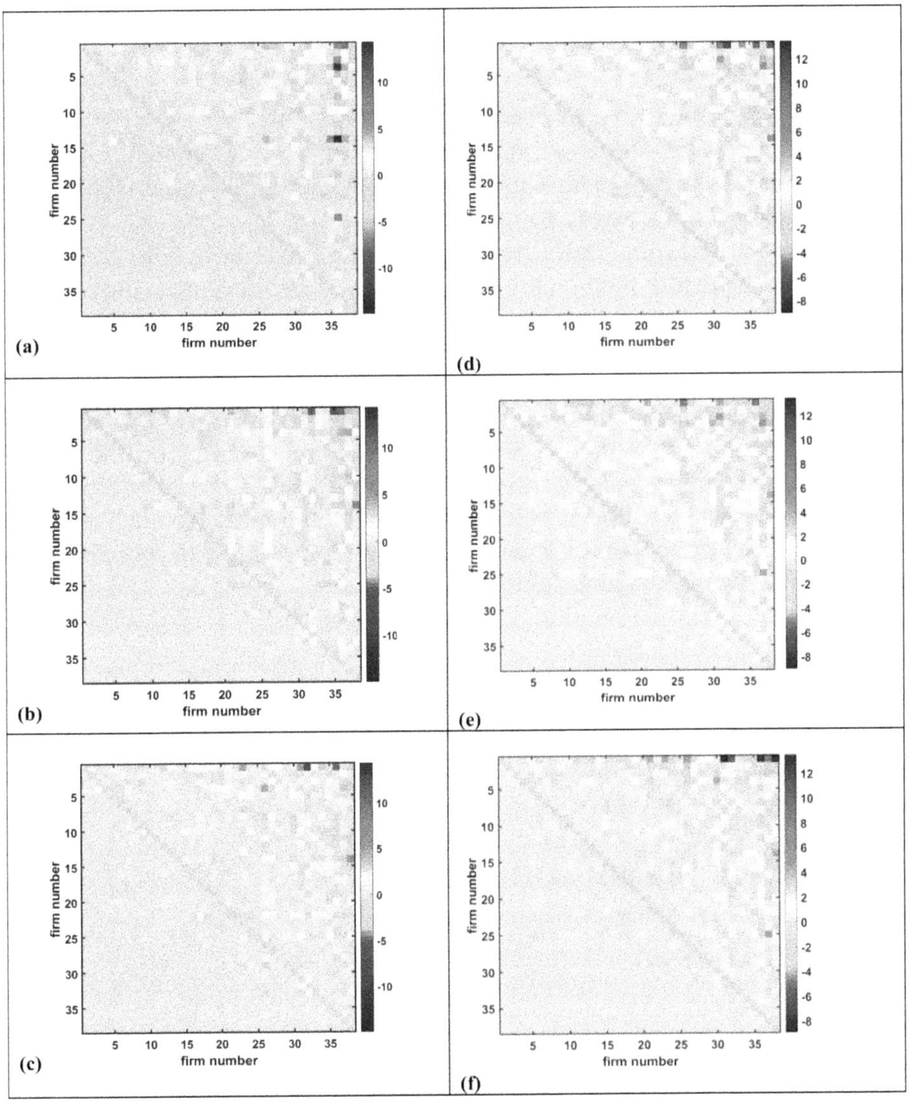

Figure 7.3. Variation of **I** with the training period. Left column, (a): $T_{\rm tr}$ = 100, (b): $T_{\rm tr}$ = 300, (c): $T_{\rm tr}$ = 500 days. Right column, (d): $T_{\rm tr}$ = 1000, (e): $T_{\rm tr}$ = 1200, (f): $T_{\rm tr}$ = 1400 days.

Therefore, for a given training period there are firms with growth rates r_i>0 and others with $r_i = 0$. Typically, half of the firms have r_i>0. The RDPME reduces for firms with $r_i = 0$ to a 'null' model which predicts that the share of company i remains constant, i.e., $x_i^{\rm pred}(t)$ = constant = $x_i(T_{\rm tr})$ for all t>$T_{\rm tr}$. An alternative estimate of r_i from equation (7.4) is by using least squares, however, this method yields larger errors.

7.5 Model validation

7.5.1 Beat the market

The random walker as standard benchmark

According to the RWH, stock market prices evolve as a random walk. As we have seen, this hypothesis is consistent with the EMH, which implies that it is impossible to 'beat the market' consistently.

We have also noticed that both the RWH and the EMH have been challenged by economists and investors who claim that stock market prices may move in trends and that the market is predictable to some degree. Indeed, there have been several economic studies and tests supporting this view (Lo 1999, 2004, Fromlet 2001, Lo *et al* 2002).

In any case, the standard model for stock prices is the *geometric Brownian motion* (Ross 2014). Therefore, here we use the discrete-time version of the geometric Brownian motion, aka *geometric random walk* (grw), as a model to compare the RDPME. Denoting by w_i the natural logarithm of the market value of company number i, it is easy to show that if it evolves as a random walk with probability p_i and 'step' s_i, then its average at time step t will be given by

$$\overline{w_i^{\mathrm{rw}}}(t) = w_i(t_0) + (2p - i1)s_i(t - t_0), \tag{7.7}$$

where $w_i(t_0)$ is its initial value at time step t_0. Then, the training period is used to estimate parameters p_i and s_i of equation (7.7) as:

$$s_i = \left|\left(v_i(T_{tr}) - v_i(1)\right)\right| / T_{tr}, \tag{7.8a}$$

$$p_i = \sum_{t=1}^{T_{tr}} \mathrm{sign}\left(w_i(t+1) - w_i(t)\right) / T_{tr}. \tag{7.8b}$$

And the grw predictions of market values for $t \leqslant T_v$ are thus obtained as:

$$\overline{\nu_i^{\mathrm{rw}}}(T_\nu) = \nu_i(T_{\mathrm{tr}})\exp\left[\left(2\sum_{t'=1}^{T_{\mathrm{tr}}-1}\frac{\mathrm{sign}(w_i(t'+1) - w_i(t'))}{T_{\mathrm{tr}} - 1} - 1\right)\frac{|(\nu_i(T_{\mathrm{tr}}) - \nu_i(1))|}{T_{\mathrm{tr}} - 1}(t - T_{\mathrm{tr}})\right]. \tag{7.9}$$

Therefore, the corresponding shares are given by:

$$x_i^{\mathrm{rw}}(t) = \frac{\nu_i(T_{\mathrm{tr}})\exp\left[\left(2\sum_{t'=1}^{T_{\mathrm{tr}}-1}\frac{\mathrm{sign}(w_i(t'+1) - w_i(t'))}{T_{\mathrm{tr}} - 1} - 1\right)\frac{|(\nu_i(T_{\mathrm{tr}}) - \nu_i(1))|}{T_{\mathrm{tr}} - 1}(t - \mathrm{tr})\right]}{\sum_{i=1}^{S}\nu_i(T_{\mathrm{tr}})\exp\left[(2\sum_{t'=1}^{T_{\mathrm{tr}}-1}\frac{\mathrm{sign}(w_i(t'+1) - w_i(t'))}{T_{\mathrm{tr}} - 1} - 1)\frac{|(\nu_i(T_{\mathrm{tr}}) - \nu_i(1))|}{T_{\mathrm{tr}} - 1}(t - T_{\mathrm{tr}})\right]}. \tag{7.10}$$

Quantifying the accuracy of predictions through the percentage errors of share trajectories

To compare accuracy for predicting future fractions x_i of the RDPME, equation (7.4), versus the one of the grw benchmark forecasting, equation (7.10), the procedure is as follows:

First, the mean absolute percentage error (MAPE) are computed over the validation period T_v, for each firm i with the two formulas. Denoting by x^{pred} predictions using either the RDPME or the grw, the MAPE for firm i is given by:

$$\mathrm{MAPE}_i = 100\frac{\sum_{t=1}^{T_v} | \, x(t)_i^{\mathrm{pred}} - x_i \, | / x_i(t)}{T_v}, \quad i = 1, 2, \ldots, S. \tag{7.11}$$

This metric quantifies the error of the predicted 'trajectory' for the share x_i of firm i between days $T_{tr} + 1$ and T_V with respect to the empirical trajectory it has followed.

Second, to obtain a global assessment of the accuracy of the method, the average of MAPE_i over all firms is taken thus producing a global MAPE denoted as GMAPE:

$$\mathrm{GMAPE} = \sum_{i=1}^{S}\frac{\mathrm{MAPE}_i}{S} = 100\sum_{i=1}^{S}\frac{\dfrac{\sum_{t=1}^{T_v} | \, x(t)_i^{\mathrm{pred}} - x_i \, | / x_i(t)}{T_v}}{S}. \tag{7.12}$$

The smaller this global metric the more accurate is the method as a whole.

7.5.2 Quantitative predictions for individual companies

Let us first analyze the accuracy of RDPME short-term and long-term predictions for each company. This is done by considering a relatively short validation period of $T_V = 50$ days (and for 14 training periods, $T_{tr} = 100, 200, \ldots, 1300, 1400$ days) and a relatively long validation period of $T_V = 500$ days (and for five training periods, $T_{tr} = 500, 600, 700, 800$ and 900 days).

Results for both validation periods $T_V = 50$ days and $T_V = 500$ days (and for the corresponding different training periods T_{tr}) are depicted in figure 7.4. In this figure it is also indicated for which firms the estimated $r_i>0$, and thus the RDPME predicts a variable share, different from the empirical share at $t = T_{tr}$. In panels (a) and (c) if the cell at (T_{tr}, i) is black it means that for this company i and this training period T_{tr} equation (7.6) produces a positive growth rate parameter r_i. Conversely, if the cell is white it means that the obtained r_i by equation (7.6) is 0 and thus the RDPME predicts a market share of this firm $x_i^{\mathrm{pred}}(t) = \text{constant} = x_i(T_{tr})$ for all $t>T_{tr}$. Indeed, roughly half of the cells in panels (a) and (c) are black, meaning that the growth rate parameter r_i produced by equation (7.6) is positive 50% of the times. For example, notice that there are two companies for which $r_i>0$ for all training periods: MSFT (company #2) and WMT (company #8). Additionally, there is one

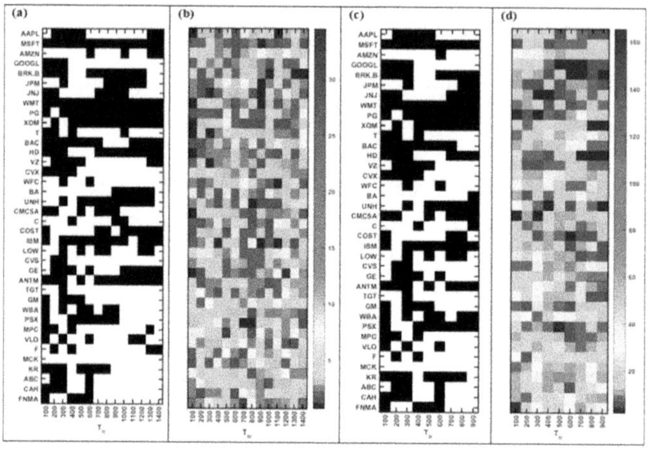

Figure 7.4. The suitability of RDPME and the accuracy of its predictions for each firm. (a) and (b): validation period $T_{\rm V} = 50$ days, training period $T_{\rm tr}+T_{\rm V} = 100, 200,\ldots,1400$. (c) and (d): validation period $T_{\rm V} = 500$ days, training period $T_{\rm tr}+T_{\rm V} = 100, 200,\ldots,900$. In panels (a) and (c) a black cell at $(T_{\rm tr},i)$ means that for this company i and this training period $T_{\rm tr}$ the obtained growth rate parameter r_i—through equation (7.6)—is positive. Conversely, a white cell means that for this company i and this training period $T_{\rm tr}$ the obtained r_i is equal to 0 and thus the RDPME method reduces to the null model, i.e., $x_i^{\rm pred}(t)=$ constant $= x_i(T_{\rm tr})$ for all $t>T_{\rm tr}$. Colored cells in panels (b) and (d) denote the corresponding MAPE for each cell $(T_{\rm tr},i)$ (equation (7.11)). Blue (red) indicates small (large) MAPE.

firm, MCK (company #34), such that $r_i = 0$ for all training periods. Colored cells in panels (b) and (d) denote the MAPE for cell $(T_{\rm tr},i)$ given by equation (7.11). Blue (red) indicates small (large) MAPE. A pattern which can be observed by direct inspection is that larger MAPE occurs for smaller cap firms. This pattern is particularly evident for $T_{\rm V} = 500$ days, where the orange-red cells occur mostly in the lower part of panel (d).

Regarding forecasting trajectories of shares of companies $x(t)$, figure 7.5 shows the empirical trajectories for the shares of 20 companies (in black) and those predicted by RDPME, equation (7.4) (red), and grw, equation (7.11) (blue), for $T_{\rm tr} = 200$ and $T_{\rm v} = 50$ days. The 20 companies are those such that equation (7.7) produces, for $T_{\rm tr} = 200$ days, an intrinsic growth rate $r_i>0$ and thus equation (7.4) leads to a share x_i that varies with time. For the other 18 firms equation (7.7) produces an intrinsic growth rate $r_i = 0$ and thus the prediction of RDPME through equation (7.4) is a horizontal line $x^{\rm pred}(t) = x_i(200)$ for all t such that $200 <t\leqslant T_{\rm V} = 50$ days (not shown in the figure). Notice that while empirical shares describe wiggled curves, those predicted by RDPME or grw do not. This difference is because the RD is a deterministic equation and the grw predictions are mean values of a stochastic process (which yields also deterministic results).

As we can see from figure 7.5, for this particular partition of $T_{\rm tr} = 200$ and $T_{\rm tr} = 50$, the RDPME model outperforms the grw benchmark (i.e., it achieves a smaller MAPE) for 16 out of 20 = 80% of these companies, it underperforms the grw for 2 in 20 = 10% and both tie for 2 in 20 = 10%. When considering all the 38

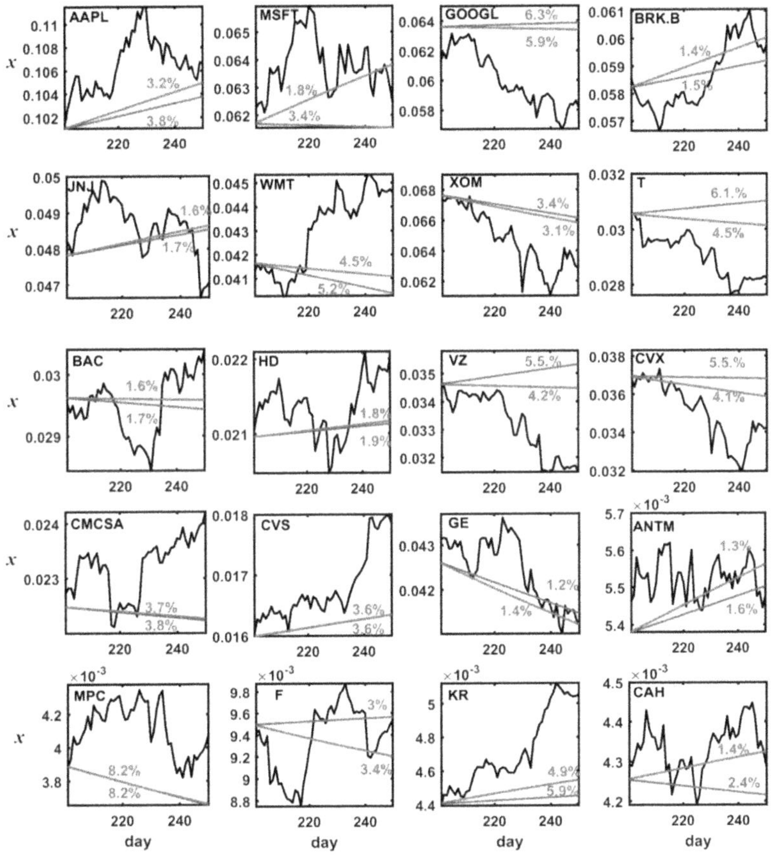

Figure 7.5. Shares for the 20 firms such that for $T_{tr} = 200$ the estimated growth rate $r_i > 0$ along $T_v = 50$ days. Empirical values (black), grw predictions (blue) and RDPME predictions (red). Percentages in each panel denote the MAPE for grw (blue) and RDPME predictions (red).

companies, including those with $r_i = 0$, these percentages become: 63%, 26% and 11%.

Figure 7.6 is the same as figure 7.5 but for a much longer training period; $T_{tr} = 700$ days. In this case there are 16 such that equation (7.7) produces an intrinsic growth rate $r_i>0$ and thus equation (7.4) leads to a share x_i that varies with time.

Now the RDPME model outperforms the grw benchmark for 10 out of 16 = 63% of the companies; these companies are MSFT, BRK.B, WMT, HD, UNH, CMSA, COST, IBM, LOW and PSX. When considering all the 38 companies, including those with $r_i = 0$, the percentage of wins becomes 53%. For the great majority of partitions of $T_{tr}-T_{tr}$, these percentages are qualitatively similar.

It is worth remarking that the MAPEs of both methods for this larger validation period are in general much larger ($\overline{\text{MAPE}}_{\text{RDPME}} = 26.1\%$, $\overline{\text{MAPE}}_{\text{grw}} = 28.1\%$). This increase in the errors is an expected result, we have to take into

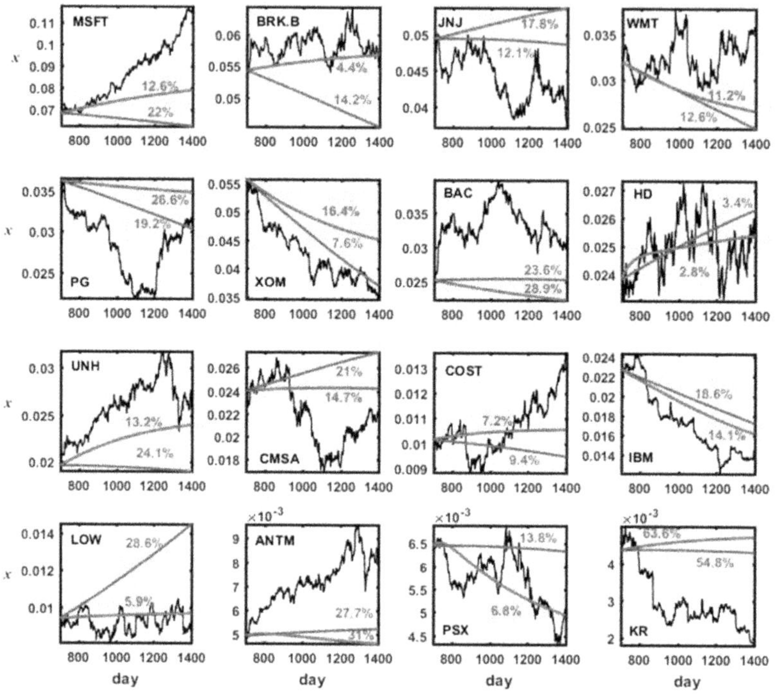

Figure 7.6. Shares for the 16 firms for which, when $T_{tr} = 700$, the estimated intrinsic growth rate $r_i>0$ and thus RDPME predicts a variable share for those companies along $T_v = 700$ days. Empirical values (black), grw predictions (blue) and RDPME predictions (red). Percentage in each panel denote the corresponding MAPE for grw predictions (blue) and RDPME predictions (red).

account that the validation period in this latter case comprises more than two years of trade.

7.5.3 Global accuracy of the RDPME method

Table 7.2 shows the GMAPE, both for the RDPME and grw, for different partitions of T_{tr} and T_v (in days).

Remarks:

i. For the great majority of partitions the RDPME yields a smaller GMAPE than the grw.
ii. For each training period the RDPME achieves always the minimum GMAPE.
iii. Most of the minimum GMAPE correspond to $T_{tr} = 200$, thus suggesting a sort of trade-off implying that very large training periods do not provide larger accuracy.
iv. As expected, for a given training period, the GMAPE grows with the validation period (i.e., the larger the forecast horizon the less accurate the predictions).

Table 7.2. GMAPE, i.e., mean absolute percentage errors (MAPE) averaged over firms produced by the Replicator Dynamics equation (RDPME) equation (7.11) compared with the geometric random walk (grw) equation (7.10) for different training and validation periods (fixed initial time day = 1). The last column corresponds to % of times the RDPME leads to a smaller relative error than the grw. Numbers in bold correspond to minimum GMAPE for a given validation period.

Validation period (days) ↓	Method	Training period (days) →													% of wins
		100	200	300	400	500	600	700	800	900	1000	1100	1200	1300	
100	RDMPE	6.2	6.6	5.4	6.4	8.8	**5.0**	7.4	5.7	6.5	7.8	6.4	8.8	7.5	76.9
	grw	6.5	7.2	5.4	6.8	9.2	5.7	7.9	5.6	7.0	6.9	6.7	9.6	7.6	23.1
200	RDMPE	11.4	9.0	9.2	10.4	11.7	**8.3**	9.8	9.0	11.0	11.2	10.1	10.8		83.3
	grw	13.3	10.0	8.8	11.1	13.5	9.9	10.6	9.5	10.2	9.4	10.9	11.9		16.7
300	RDMPE	13.6	**10.9**	12.9	13.1	13.9	11.5	13.0	13.3	15.3	13.0	12.8			63.6
	grw	18.5	12.0	12.3	15.0	16.7	13.9	14.4	12.9	13.6	11.6	13.6			36.4
400	RDMPE	15.6	**12.4**	16.1	15.0	16.3	15.1	16.8	17.8	17.8	14.5				60
	grw	23.6	13.3	16.1	18.2	20.4	18.4	18.0	16.7	16.2	13.9				40
500	RDMPE	17.5	**13.5**	19.1	16.7	19.4	18.6	20.7	20.8	19.8					77.8
	grw	30.0	14.4	19.4	21.6	25.0	22.9	21.8	19.9	18.8					22.2
600	RDMPE	19.0	**14.8**	22.2	19.1	22.5	22.2	23.6	23.3						87.5
	grw	36.6	15.7	22.7	25.8	29.6	27.4	25.0	23.1						12.5
700	RDMPE	20.2	**16.4**	25.7	21.6	25.8	25.0	26.1							100
	grw	41.9	17.4	26.4	30.1	34.6	31.6	28.1							0
% of wins RDEMPE		100	100	57.1	100	100	100	100	16.7	16.7	0	100	100	100	

RDPME: replicator dynamics, grw: geometric random walk

7.6 Conclusion: balance, caveats, extensions and improvements

Natural selection in biology rests on a rigorous explanatory power, including quantitative checkable predictions. For example, we have seen in chapter 4 how a simple modelling is able to reproduce the observed frequency variations of COVID-19 variants. On the other hand, its economic counterpart is often an example of the truism that surviving firms are efficient because only the efficient survive (Knudsen 2002). In particular, the model of RD provides important insights in the evolution of markets but has not met with much empirical support (Cantner *et al* 2019).

Here we attempted to fill that gap by providing quantitative testable predictions that hopefully will contribute toward making the replicator dynamics more than just a theoretically elegant model but also a relevant economic quantitative tool regarding empirical data. With this aim we resorted to the maximum entropy principle, a fruitful approach used in several different fields to make least biased inferences with incomplete information, to infer from empirical financial time series the payoff matrix of the RD equation.

The main finding was that the resulting RDPME method outperforms the neutral benchmark of the geometric random walk (grw) in predicting future shares x_i for most of the companies and for most of the partitions in training and validation periods. It is worth mentioning that the relative errors of the RDPME are in general only modestly smaller than those produced by the grw. Hence, perhaps this is a sign that financial markets need to be explained in terms of truly complex systems modelling (Kuhlmann 2014).

Let us conclude this chapter by reviewing some caveats, extensions and possible future improvements. A general intrinsic limitation of the RD as a forecasting tool for stock prices is that it works with shares or frequencies as variables so it can predict market shares rather than market caps or the price of a stock. One way to overcome this drawback is by introducing an additional variable M for the total market value of the market (i.e., the sum of all market caps of the S firms considered). This variable is shown in figure 7.7 and exhibits a clear growing trend. Therefore we can fit the trajectory of $M(t)$ for a given T_{tr} using least squares and obtain a predicted total cap for the validation period T_{v}. Again, how to do this is not

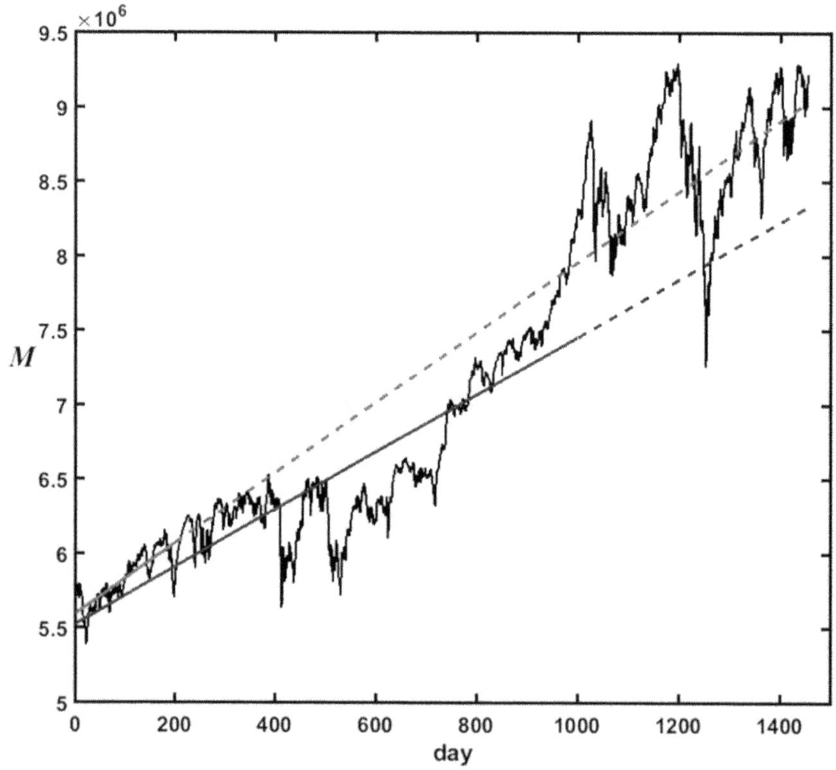

Figure 7.7. Time evolution of the aggregated market cap of the 38 companies $M(t)$ along the considered period of 1455 days. Empirical values (black) and RDPME predictions (dashed lines) through linear regression, computed from the first T_{tr} days, for the remaining $T_{\mathrm{v}} = 1455 - T_{\mathrm{tr}}$ days; for $T_{\mathrm{tr}} = 1000$ (blue) and $T_{\mathrm{tr}} = 200$ days (red).

trivial and involves a sort of handy craft. If for example we choose $T_{tr} = 1000$ days, the least squares prediction (blue line) clearly underestimates the total market cap for $t = 1001,\ldots,1455$ days. Surprisingly, a substantially better prediction for the latter part of the time series is achieved for this system using a much shorter $T_{tr} = 200$ days (red line).

Another limitation of the RD equation in describing the evolution of competing phenotypes is that it does not incorporate mutation and so is not able to innovate new types of phenotypes. To overcome such a drawback the RD equation is usually extended to the replicator–mutator (RM) equation (Nowak 2006), i.e., the RD equation with mutations. With the introduction of changes by random mutations of the set of phenotypes it includes a source of innovation lacking in the RD. The RM would allow one to model changes in the list of top firms (actually this list usually changes from one year to the next).

There is also a caveat regarding the set of selected firms: they are the largest (in revenues) US listed companies in the NYSE. But there are also some non-US companies operating in the US with larger revenues than some of the companies in this set. We used this criterion of US based firms to select the 'community' because it seemed more clear cut and due to the global dominance of US companies (Ross 2020).

Appendix A: A metric to measure the pace of change of the payoff matrix

As mentioned, the **I** matrix changes with the training period. From simple visual inspection it seems that **I** changes faster from lower values of T_{tr}. In the appendix at the end of this chapter we propose a metric. Therefore, it is important to measure the variation rate of **M** because the faster the pace the shorter the validation period. With this aim we introduced the following metric for comparing differences between the non-diagonal parts of **M** at two times, t and t':[5]

$$\delta \boldsymbol{M}(t, t') \equiv \left\langle \left| M_{ij}(t) - M_{ij} - (t') \right|_{i \neq j} \right\rangle = \frac{\sum_{i}^{S} \sum_{j \neq i}^{S} \left| M_{ij}(t) - M_{ij} - (t') \right|}{(S-1)^2}, \tag{A.1}$$

where we denoted by the angle brackets $\langle\rangle$ averages over all $(S-1)\times(S-1)$ non-diagonal matrix elements. We take the absolute value of variations of each matrix element to avoid cancellations when performing these averages.

Figure A.1 shows the variation of $\delta \boldsymbol{M}(T_{tr}, T_{tr} + 100)$, given by equation (A.1), versus the time t measured in days. That is, the first point corresponds to the difference between a training period $T_{tr} = 100$ and a time $t = T_{tr} + 100 = 200$, the second point corresponds to the difference between a training period $T_{tr} = 200$ and a time $t = T_{tr} + 100 = 300$, and so on. Notice that $\delta \boldsymbol{M}(T_{tr}, T_{tr} + 100)$ first decreases quickly and then after $T_{tr} = 600$ seems to stabilize to a value around 0.12.

[5] Alternatively, the Euclidean norm can be used. I checked it produces similar results although they are less clear-cut.

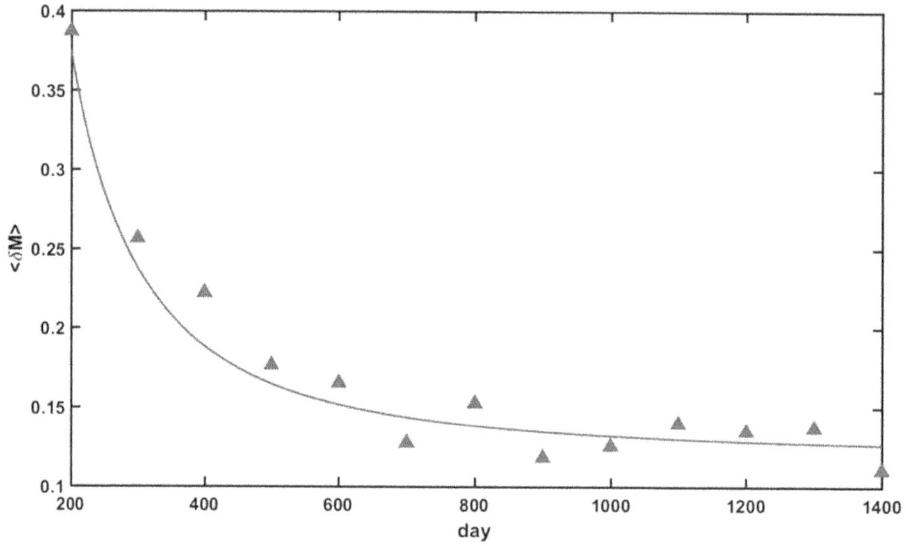

Figure A.1. Variation with time (measured in days) of **M**, for intervals of 100 days and training periods of 100, 200, …,1300 days (triangles) and the corresponding fit (filled curve).

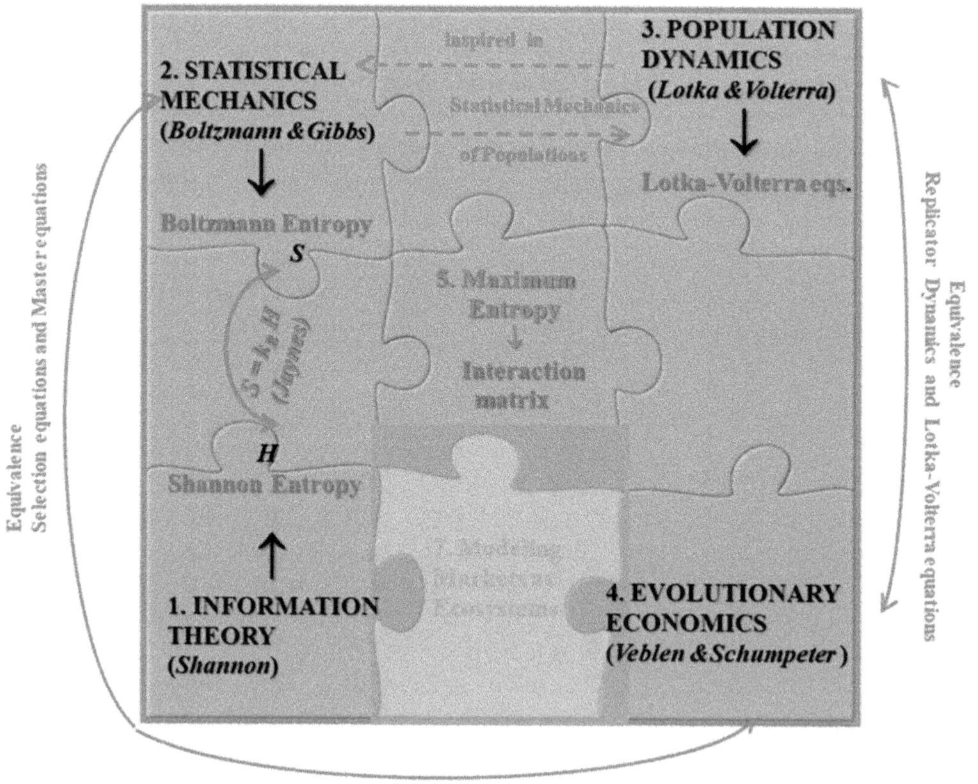

References

Astley W G and Fombrun C J 1983 Collective strategy: social ecology of organizational environments *Acad. Manage. Rev.* **8** 576–87

Bachelier L 1900 Théorie de la spéculation *Ann. Sci. Ec. Norm. Super.* **17** 21–86

Boyte-White C 2020 Market capitalization versus market value: what's the difference? *Investopedia* https://investopedia.com/ask/answers/122314/what-difference-between-market-capitalization-and-market-value.asp accessed 11 January 2022

Cantner U, Savin I and Vannuccini S 2019 Replicator dynamics in value chains: explaining some puzzles of market selection *Ind. Corp. Change* **28** 589–611

Cantner U 2017 Structural change in heterogeneous actor populations *Foundations of Economic Change: A Schumpeterian View on Behaviour, Interaction and Aggregate Outcomes* ed A Pyka and U Cantner (Berlin: Springer)

Case J 2008 *Competition: The Birth of a New Science* (New York: Hill and Wang)

Chan L K C, Karceski J J and Lakonishok J 1999 *On Portfolio Optimization: Forecasting Covariances and Choosing the Risk Model* NBER Working paper No. w7039, Available at SSRN:https://ssrn.com/abstract=156690

Chen M J 2002 Transcending paradox: the Chinese 'middle way' perspective *Asian-Pac. J. Manage.* **19** 179–99

Clarke-Hill C, Li H and Davies B 2003 The paradox of co-operation and competition in strategic alliances: towards a multi-paradigm approach *Manage. Res. News* **26** 1–20

Cootner P H 1964 *The Random Character of Stock Market Prices* (Cambridge, MA: MIT Press)

Czakon W 2010 Emerging coopetition: an empirical investigation of coopetition as inter-organizational relationship instability *Coopetition. Winning Strategies for the 21st Century* ed S Yami, S Castaldo, G Dagnino and F Le Roy (Cheltenham: Edward Elgar Publishing) pp 58–73

Dhir R 2021 Efficient market hypothesis: is the stock market efficient? *Investopedia* https://investopedia.com/articles/basics/04/022004.asp accessed 11 October 2021

Dosi G and Nelson R R 1994 An introduction to evolutionary theories in economics *J. Evol. Econ* **4** 153–72

Downey L 2021 Efficient market hypothesis (EMH) *Investopedia* https://investopedia.com/terms/e/efficientmarkethypothesis.asp accessed 4 October 2021

Dyer J and Singh H 1998 The relational view: cooperative strategy and sources of interorganizational competitive advantage *Acad. Manage. Rev.* **24** 660–79

Einstein A 1905 Über die von der molekularkinetischen Theorie der Wärme geforderte Bewegung von in ruhenden Flüssigkeiten suspendierten Teilchen' [On the movement of small particles suspended in stationary liquids required by the molecular-kinetic theory of heat] *Ann. Phys.* **322** 549–60

Emary C and Fort H 2021 Markets as ecological networks: inferring interactions and identifying communities *J. Complex Netw.* **9** 1–17

Engle R F 2009 *Anticipating Correlations: A New Paradigm for Risk Management* (Princeton, NJ: Princeton University Press)

Fama E F 1965 Random walks in stock market prices *Financ. Anal. J.* **21** 55–9

Fama E F 1970 Efficient capital markets: a review of theory and empirical work *J. Finance* **25** 383

Fortune 2018 https://fortune.com/fortune500/2018/

Fortune 2019 https://fortune.com/fortune500/2019/

Fromlet H 2001 Behavioral finance—theory and practical application *Bus. Econ.* **36** 63–9

Knudsen T 2002 Economic selection theory *J. Evol. Econ.* **12** 443–70

Kuhlmann M 2014 Explaining financial markets in terms of complex systems *Phil. Sci.* **81** 1117–30

Lakka S, Michalakelis C, Varoutas D A and Martakos D 2013 Competitive dynamics in the operating systems market: modeling and policy implications *Technol. Forecast. Soc. Change* **80** 88–105

Liberto D 2021 Adaptive market hypothesis (AMH) *Investopedia* https://investopedia.com/terms/a/adaptive-market-hypothesis.asp accessed 4 October 2021

Lo A 1999 *A Non-Random Walk Down Wall Street* (Princeton, NJ: Princeton University Press)

Lo A, Mackinlay W and Craig A 2002 *A Non-Random Walk Down Wall Street* 5th edn (Princeton, NJ: Princeton University Press) pp 4–47

Lo A W 2004 The efficient market hypothesis: market efficiency from an evolutionary perspective *J. Portfolio Manage.* **30** 15–29

Malkiel B G 1973 *A Random Walk Down Wall Street* 6th edn (New York: W.W. Norton & Company, Inc.)

Mazzucato M 2000 *Firm Size, Innovation, and Market Structure: The Evolution of Industry Concentration and Instability* (Cheltenham: Edward Elgar Publishing)

Moore J F 1993a Predators and prey: a new ecology of competition *Harvard Business Review* May/June 1993

Moore J F 1993b *The Death of Competition: Leadership and Strategy in the Age of Business Ecosystems* (New York: Harper Paperbacks)

Morningstar 2019 *Morningstar's Active/Passive Barometer* August 2019

Nelson R R and Winter S G 1982 *An Evolutionary Theory of Economic Change* (Cambridge, MA: Harvard University Press)

Nowak M A 2006 *Evolutionary Dynamics: Exploring the Equations of Life* (Cambridge, MA: Belknap Press) pp 272–73

NYSE 2021 https://nyse.com/market-cap accessed 17 September 2021

Oliver A L and Ebers M 1998 Networking network studies: an analysis of conceptual configurations in the study of inter-organizational relationships *Organ. Stud.* **19** 549–83

Pearson K 1905 The problem of the random walk *Nature* **72** 294–342

Pfizer 2020 *Pfizer and Biontech Announce Further Details on Collaboration to Accelerate Global Covid-19 Vaccine Development* https://investors.pfizer.com/investor-news/press-release-details/2020/Pfizer-and-BioNTech-Announce-Further-Details-on-Collaboration-to-Accelerate-Global-COVID-19-Vaccine-Development/default.aspx accessed 18 June 2021

Pfizer 2021 *Pfizer and BioNTech to Provide 500 Million Doses of COVID-19 Vaccine to U.S. Government for Donation to Poorest Nations* https://pfizer.com/news/press-release/press-release-detail/pfizer-and-biontech-provide-500-million-doses-covid-19 accessed 18 June 2021

Porter M E 1980 *Competitive Strategy* (New York: Free Press)

Protos 2021 *Crypto Trading Hamster Outperforms Bitcoin, Warren Buffett, Cathie Wood* https://protos.com/crypto-trading-hamster-goxx-outperforms-bitcoin-buffett-wood/

Quandl 2019 *Core US Fundamentals Dataset* Available at http://www.quandl.com/databases/SF1

Rayleigh L 1905 The problem of the random walk *Nature* **72** 318

Regnault J 1863 *Calcul des Chances et Philosophie de la Bourse* (London: Forgotten Books) Classic Reprint Series, 4 March 2018

Ross S M 2014 Variations on Brownian motion *Introduction to Probability Models* 11th edn (Amsterdam: Elsevier) pp 612–14

Ross J 2020 *The Dominance of U.S. Companies in Global Markets* https://visualcapitalist.com/us-companies-global-markets/ accessed 18 November 2021

Rumelt R P 1991 How much does industry matter? *Strateg. Manage. J.* **12** 167–85

Samuelson P 1965a Rational theory of warrant pricing *Ind. Manage. Rev.* **6** 13–39

Samuelson P 1965b Proof that properly anticipated prices fluctuate randomly *Ind. Manage. Rev.* **6** 41–9

Shmueli G and Lichtendahl K C 2016 *Practical Time Series Forecasting with R: A Hands-on Guide* 2nd edn (Green Cove Springs, FL: Axelrod Schnall Publishers)

Toynbee A J 1946 *A Study of History, Vol. 1: Abridgement of Volumes I–VI* New Edition 1999 (Oxford: Oxford University Press)

Vasicek O A and McQuown J A 1972 The efficient market model *Financ. Anal. J.* **28** 71–84

Yami S *et al* 2010 Introduction—coopetition strategies: towards a new form of inter-organizational dynamics? *Coopetition. Winning Strategies for the 21st Century* ed S Yami, S Castaldo, G Dagnino and F Le Roy (Cheltenham: Edward Elgar Publishing) pp 58–73

IOP Publishing

Glossary

Italic words refer to other entries of the glossary.

ADIABATICITY or QUASISTATIC EVOLUTION A slow process in which all time derivatives are very small.

ALTERNATIVE STABLE STATES (ASS) There are ecosystems that for given environmental conditions (sets of unique biotic and abiotic conditions) can exist in different alternative 'states'. These alternative states are non-transitory and therefore considered stable over ecologically relevant timescales. Interestingly, ecosystems may transition from one stable state to another, in what is known as a state shift or regime shift, when perturbed. Alternative stable state theory suggests that discrete states are separated by ecological thresholds, in contrast to ecosystems which change smoothly and continuously along an environmental gradient. Mathematically this corresponds to nonlinear systems in which a variation of certain parameters can cause different types of *bifurcations*. Stable equilibria, corresponding to such alternative stable states, can appear or disappear, leading to a shift from one stable state to a radically different one, from which recovery is exceedingly difficult. Pairs of stable equilibria (*attractor*) are separated by an unstable equilibrium. Each attractor has its own *basin of attraction*, i.e., the sets of points from which a *dynamical system* spontaneously moves to this particular attractor. The unstable equilibrium marks the border between the basins of attraction of both stable states.

ATTRACTOR In *dynamical systems*, an attractor is a set of numerical values toward which a system tends to evolve, for a wide variety of starting conditions of

doi:10.1088/978-0-7503-3931-5ch8

the system. System values that get close enough to the attractor values remain close even if slightly disturbed, in such a way that all trajectories not contained in that region will eventually wind up in the region. An attractor may be a point or a cycle that is an equilibrium and generates transients that return to the equilibrium state after perturbation. It may also be an attractive region that has no individual equilibrium points or cycles (a chaotic or strange attractor).

AUTONOMOUS DYNAMICAL SYSTEM A system of ordinary differential equations which does not explicitly depend on the independent variable. When the independent variable is time, they are also called time-invariant systems.

BASIN OF ATTRACTION For each attractor, its basin of attraction is the set of initial conditions leading to long-time behavior that approaches that attractor. That is, the collection of points that converge on a particular attractor.

BIFURCATION A bifurcation occurs when a small smooth change made to the parameter values (the 'bifurcation' parameters) of a system causes a sudden 'qualitative' or topological change in its behavior.

BIFURCATION DIAGRAM A graph of the attractors of a system as a function of some parameter (the 'bifurcation' parameter). It shows the values visited or approached asymptotically (fixed points, periodic orbits, or chaotic attractors) of a system as a function of this bifurcation parameter in the system.

BIFURCATION, LOCAL A local bifurcation occurs when a parameter change causes the stability of an equilibrium (or fixed point) to change. In continuous systems, this corresponds to the real part of an eigenvalue of an equilibrium passing through zero. In discrete systems (those described by maps rather than ODEs), this corresponds to a fixed point having an eigenvalue with modulus equal to one. In both cases, the equilibrium is **non-hyperbolic** (at least the real part of one eigenvalue becomes zero) at the bifurcation point. The topological changes in the phase portrait of the system can be confined to arbitrarily small neighborhoods of the bifurcating fixed points by moving the bifurcation parameter close to the bifurcation point (hence 'local'). By contrast, global bifurcations cannot be revealed by eigenvalue degeneracy.

BIFURCATION, NORMAL FORM OF In mathematics, the normal form of a *dynamical system* is a simplified form that can be useful in determining the system's behavior. Normal forms are often used for determining local bifurcations in a system. All systems exhibiting a certain type of bifurcation are said to be locally (around the equilibrium) topologically equivalent to the normal form of the bifurcation.

BIFURCATION POINT A point of structural instability in which a single equilibrium condition is split into two.

CANONICAL ENSEMBLE The canonical ensemble of a gas is the *statistical ensemble* that represents the possible microstates of this gas—described by a point in the phase space $[Q,P]$ of all coordinates and momenta of molecules $\{q_i, p_i\}$—in thermal equilibrium with a heat reservoir or heat bath at a fixed temperature T. This heat reservoir is assumed to have an infinitely large heat capacity, so that, irrespective of the energy exchange between the gas and the reservoir the temperature T can be maintained. The gas, of N molecules and occupying a volume V, is assumed surrounded by a *diathermal* membrane, i.e., a wall that allows heat transfer but does not allow transfer of matter across it. Additionally the membrane is rigid. Thus a canonical ensemble is a collection of systems characterized by the same values of N, V and T, which define the macrostate of the gas. Since in the canonical ensemble the gas can exchange energy (in the form of heat) with the heat reservoir, the energy E of the gas is variable; in principle, it can take values anywhere between zero and infinity. The probabilistic weights for the energy are given by the canonical probability distribution $P(E) \propto \exp\left(\frac{E}{k_B T}\right)$ (k_B is the Boltzmann constant).

CARRYING CAPACITY The maximum attainable size of a population, usually symbolized as K.

CATASTROPHE or IMPERFECT BIFURCATION A catastrophe occurs when the stability of an equilibrium breaks down, causing the system to jump into another state. This jump could be truly catastrophic for the equilibrium of a bridge or a building or a species that extinguishes. Catastrophes can be also regarded as *imperfect bifurcations*, often described by the addition of an imperfection parameter to the *normal form of a bifurcation.*

CHARACTERISTIC POLYNOMIAL The characteristic polynomial of a square matrix is a polynomial which has the eigenvalues as roots. Let A be an $n \times n$ matrix. The characteristic polynomial of A, denoted by $p_A(\lambda)$ is the polynomial defined by equating the determinant of $\mathbf{A} - \lambda\mathbf{I}$ (where $\mathbf{I}$ is the identity matrix) to 0.

COMPETITIVE EXCLUSION PRINCIPLE sometimes referred to as Gause's law, is the proposition that two species competing for the same limiting resource cannot coexist at constant population values. When one species has even the slightest advantage over another, the one with the advantage will dominate in the long term. This result can be derived from the Lotka–Volterra competition equations: if interspecific competition between two species is sufficiently large, the equilibrium of both species coexisting is unstable.

DENSITY DEPENDENCE The condition in which the rate at which a population increases or decreases is a function of its density (in contrast with density independence).

DYNAMICAL SYSTEM A means of describing how one state develops into another state over the course of time in terms of a system of equations. These equations describe the time dependence of a point's position in its ambient (geometrical) space. *Dynamical systems theory* brings a qualitative and geometrical approach to the analysis of ordinary differential equations (ODE), addressing the existence, stability, and global behavior of sets of solutions, rather than seeking exact or approximate expressions for individual solutions.

EQUILIBRIUM POINT The value of a variable that does not change under the rules of a *dynamical system*. An equilibrium point may be stable (in which case it is commonly referred to as an *attractor*) or unstable (in which case it is commonly referred to as a *repeller*).

ENSEMBLE (STATISTICAL ENSEMBLE) In statistical mechanics, an ensemble (also statistical ensemble) is an idealization consisting of a large number of virtual copies (sometimes infinitely many) of a physical system, each of which represents a possible microstate that the system might be in.

ERGODICITY, ERGODIC THEOREM In mathematics, ergodicity expresses the idea that a point representing the state of a moving system, either a *dynamical system* or a stochastic process, will eventually visit all parts of the space that the system moves in. This implies that the average behavior of the system can be deduced from the trajectory of a 'typical' point. The ergodic theorem or ergodic hypothesis was first introduced in statistical mechanics by Boltzmann (1871). According to this hypothesis, the trajectory of a representative point (corresponding to a microstate) passes, in the course of time, through each and every point of the relevant region of the phase space. This statement was later relaxed to the so-called quasiergodic hypothesis, according to which the trajectory of a representative point traverses, in the course of time, any neighborhood of any point of the relevant region. Ergodicity thus warrants that the time averages and the *ensemble* averages are the same. Intuitively, since the trajectory $[Q(t),P(t)]$ covers the whole available phase space, the average of any property $O([Q(t),P(t)])$ over time is the same as the average of O over the collection of microstates of the *ensemble*.

EULER'S CONSTANT The base of natural logarithms, normally symbolized by a lowercase e, approximately 2.7183.

FERROMAGNETISM The basic mechanism by which certain materials, such as iron and nickel, form permanent magnets. Microscopically the ferromagnetism is explained in terms of the electrons contained in the material. Specifically, one of the fundamental properties of an electron is that it has a magnetic dipole moment, i.e., it behaves as a tiny magnet. When these tiny magnetic dipoles are aligned in the same direction, their individual magnetic fields add together to create a measurable macroscopic magnetic field.

FIRST LAW OF THERMODYNAMICS This law is a version of the law of conservation of energy, adapted for thermodynamic processes. The law of conservation of energy states that the total energy of an isolated system—i.e., a system enclosed by rigid immovable walls through which neither matter nor energy can pass—is constant. Energy can be transformed from one form to another, but can be neither created nor destroyed. In cases in which the system can exchange energy with its surroundings, either as heat Q or as thermodynamic work W but not matter, the first law is written as $\Delta U = Q + W$, where ΔU denotes the change in the internal energy of such system. A positive (negative) Q corresponds to a quantity of energy supplied to (by) the system as heat; likewise a positive (negative) W corresponds to the amount of thermodynamic work done on (by) the system by (on) its surroundings.

FREQUENCY-DEPENDENT SELECTION In biology it corresponds to the situation where fitness is dependent upon the frequency of a phenotype or genotype in a population. In the case of negative (positive) frequency-dependent selection, fitness of a phenotype or genotype increases as its frequency in a population decreases (increases). More generally, frequency-dependent selection occurs when the fitness of a type depends on whether it is rare or common, that is, the fitness of a type by the frequencies of other types in the community.

FUNCTIONAL RESPONSE In ecology a functional response is the intake rate of a consumer as a function of food density.

	That is, in consumer–resource (predator–prey) equations, the function that stipulates how the per capita consumption rate (or predation rate) changes with changes in resource density.
GAUSE PRINCIPLE	See **COMPETITIVE EXCLUSION PRINCIPLE.**
HAMILTONIAN	The mathematical descriptor for the energy of a given interaction. The total Hamiltonian describes all energies of all the interactions that affect the system.
HUTCHINSON'S MULTIDIMENSIONAL NICHE	The dimensions of the Hutchinsonian niche can be thought as environmental conditions and resources (e.g., light, nutrients, etc), that define the requirements of individuals of a species for its population to persist.
HYPERBOLIC EQUILIBRIUM	An equilibrium $\mathbf{x}^*$ is hyperbolic if no eigenvalue of the *linearization* has real part equal to zero, i.e., hyperbolic equilibrium implies that $\mathrm{Re}(\lambda_1) \neq 0$ and $\mathrm{Re}(\lambda_2) \neq 0$.
INTRASPECIFIC COMPETITION	The competitive interaction among individuals in the same population.
INTRINSIC RATE OF NATURAL INCREASE	The growth of a population under the theoretical state of extremely low population density, usually symbolized as *r*.
ISING MODEL	The Ising model aka Lenz–Ising model aka Ising–Lenz model, named after the physicists Ernst Ising and Wilhelm Lenz, is a *lattice model* originally devised to model ferromagnetism in statistical mechanics. The model consists of discrete variables that represent magnetic dipole moments of atomic 'spins' that can be in one of two states (+1 or −1). The spins are arranged in a graph, usually a lattice (where the local structure repeats periodically in all directions), allowing each spin to interact with its neighbors. Neighboring spins which are parallel have a lower energy than those which are anti-parallel. Hence, the system tends to the lowest energy (all spins parallel). However, heat disturbs this tendency, thus creating the possibility of different structural phases. The model allows the identification of phase transitions as a simplified model of reality. The two-dimensional square-lattice Ising model is one of the simplest statistical models to show a phase transition and, simultaneously, it is one of few exactly solvable models where we can actually compute thermodynamic quantities. The Ising model is simple, yet it can be applied to a surprising number of different systems in physics and in other disciplines like neuroscience, genomics, ecology, mathematical finance, etc.

ISOCLINE or NULLCLINE
In population dynamics, the term isocline refers to the set of population sizes at which the rate of change for one population in a pair of interacting populations is zero. More generally, for a *dynamical system*, the set of all points for which one of the variables does not change, so that the time derivative is equal to zero.

LATTICE MODEL
In physics, a lattice model is a physical model that is defined on a lattice, as opposed to the continuum of space. Lattice models originally occurred in the context of condensed matter physics, where the atoms of a crystal automatically form a lattice. Currently, lattice models are quite popular in theoretical physics, for two main reasons. Firstly, some models, which are very difficult to treat in continuum space, become exactly solvable in the lattice. Secondly, lattice models are also ideal for study by the methods of computational physics, as the discretization of any continuum model automatically turns it into a lattice model. An example of a lattice model in condensed matter physics is the *Ising model.*

LIMIT CYCLE
In *dynamical systems* of at least two dimensions (e.g., two species in an ecosystem) a limit cycle C is a closed trajectory in phase space which is called stable, unstable, or semi-stable according to whether the nearby curves spiral towards C, away from C, or both. In other words, a limit cycle is a closed trajectory in phase space having the property that at least one other trajectory spirals into it either as time approaches infinity or as time approaches negative infinity. Such behavior is exhibited in some nonlinear systems. Limit cycles have been used to model the behavior of a great many real-world oscillatory systems. The study of limit cycles was initiated by the French mathematician Henri Poincaré.

LINEARIZATION
Finding the linear approximation to a mathematical function at a given point is called linearization. The linear approximation of a function is the first order Taylor expansion around the point of interest. In the study of *dynamical systems*, linearization is a method for assessing the local stability of an equilibrium point of a system of nonlinear differential equations or discrete dynamical systems.

LOGISTIC EQUATION
Sometimes called the Verhulst model or logistic growth curve, is a model of population growth first published by Pierre Verhulst (1845). The model is continuous in time, but a modification of the continuous equation to a discrete quadratic

recurrence equation known as the logistic map is also widely used.

LOGISTIC POPULATION GROWTH Population growth that appears qualitatively exponential at low population density but approaches an asymptote as the population becomes larger; population growth that follows the logistic equation.

MACROECOLOGY The subfield of ecology that deals with the study of relationships between organisms and their environment at large spatial scales to characterize and explain statistical patterns of abundance, distribution and diversity.

MALTHUS EQUATION The simplest population equation describing an exponentially growing population, introduced by Thomas R. Malthus in 1798.

MARKOV CHAIN A stochastic model describing a sequence of possible events in which the probability of each event depends only on the state attained in the previous event.

MASTER EQUATION A differential equation that describes the evolution of the probabilities for a Markov process of a system that jumps from one state to other state in a continuous time. That is, it is the continuous time version of the recurrence relations for *Markov chains*.

MEAN-FIELD APPROXIMATION (MFA) In physics and probability theory, the mean-field approximation consists in approximating a random (stochastic) model by a simpler model that results from averaging over degrees of freedom. Such models consider many individual components that interact with each other. The effect of all the other individuals on any given individual is approximated by a single averaged effect, thus reducing a **many-body problem** to a **one-body problem**.

METASTABILITY In physics, metastability is a stable state of a dynamical system other than the system's state of least energy. In isolation the state of least energy is the only one the system will inhabit for an indefinite length of time, until more external energy is added to the system. That is, the system will spontaneously leave any other state (of higher energy) to eventually return (after a sequence of transitions) to the least energetic state. A ball resting in a hollow on a slope is a simple example of metastability. If the ball is only slightly pushed, it will settle back into its hollow, but a stronger push may start the ball rolling down the slope.

NICHE The ecological niche is the match of a species to a specific environmental condition. It describes how an organism or population responds to the distribution of resources required for thriving and competitors over these resources.

NULLCLINE = ISOCLINE

ORDINARY DIFFERENTIAL EQUATIONS (ODE) A differential equation containing one or more functions of one independent variable and the derivatives of those functions. The term *ordinary* is used in contrast with the term partial differential equation (PDE) which may be with respect to *more than* one independent variable.

ONE-DIMENSIONAL MAP A function f that projects a single variable x_t through discrete time t, $x_{t+1} = f(x_t)$. For example, in the logistic map $f(x_t) = rx_t(1 - x_t)$.

PAIRWISE MAXIMUM-ENTROPY PROBABILITY MODEL A model based on the principle of maximum entropy, which states that the probability distribution (i.e., the model) that best represents one's current state of knowledge about a system is the one with the largest entropy (or uncertainty) subject to known constraints on observed data. In the case of pairwise PMEPM, these constraints are the means and covariances of the model variables. Also referred to as PME (pairwise maximum entropy) modelling.

PARTIAL DIFFERENTIAL EQUATION (PDE) A differential equation that contains several unknown variables and their partial derivatives (i.e., the derivative with respect to one of those variables, with the others held constant). PDEs are used to formulate problems involving functions of several variables, typically space coordinates and time. A special case is ordinary differential equations (ODEs), which deal with functions of a single variable and their derivatives.

PHASE SPACE In *dynamical system* theory, a phase space is a space in which all possible states of a system are represented, each one as a unique point in the phase space. For mechanical systems, the phase space usually consists of all possible values of position variables $\{q_1, q_2, \ldots, q_{3N}\}$ and momentum variables$\{p_1, p_2, \ldots, p_{3N}\}$.

POPULATION A group of individual items. In the context of population ecology, a population is a group of individual living organisms.

REPELLER A point or cycle that is theoretically an equilibrium but generates transients that deviate from the equilibrium position when perturbed.

SAD: SPECIES-ABUNDANCE DISTRIBUTION Also called the relative abundance distribution (RAD) or species abundance distribution. In ecology, it describes the relationship between the number of species observed in a field study as a function of their observed abundance.

SAR: SPECIES–AREA RELATIONSHIP A metric often studied in macroecology. It describes the relationship between the area of a habitat, or of part of a habitat, and the number of species found within that area. The idea is that larger areas tend to contain larger numbers of species.

SEPARATRIX The separatrix is the boundary between two basins of attraction.

SIMULATION A numerical simulation is a calculation that is run on a computer following a program that implements a mathematical model for a physical system. Numerical simulations are required to study the behavior of systems whose mathematical models are too complex to provide analytical solutions, as in most nonlinear systems.

STRANGE ATTRACTOR A region of space that attracts all trajectories but contains no attractive points or cycles. Indeed, an attractor is called strange if it has a fractal structure, i.e., strange attractors have detailed structure on all scales of magnification. This is often the case when the dynamics on it are chaotic (but strange nonchaotic attractors also exist). If a strange attractor is chaotic, exhibiting sensitive dependence on initial conditions, then any two arbitrarily close alternative initial points on the attractor, after any of various numbers of iterations, will lead to points that are arbitrarily far apart (subject to the confines of the attractor), and after any of various other numbers of iterations will lead to points that are arbitrarily close together.

STRUCTURAL STABILITY A higher-level stability concept in which the qualitative nature of a *dynamical system* is unchanged when the parameters of the system are varied. This fundamental property of a *dynamical system* means that the qualitative behavior of the trajectories is unaffected by small perturbations of the parameters.

STRUCTURED MODELS Models that do not assume that all individuals in the population are identical. For example, models can be spatially structured (spatial heterogeneous environment), age-structured or sex-structured.

VECTOR FIELD The set of vectors that determine the behavior of a *dynamical system*.

VAN DER WAALS EQUATION OF STATE In thermodynamics, statistical mechanics or chemistry, the Van der Waals equation of state is an equation of state which extends the ideal gas law to include the effects of interaction between molecules of a gas, as well as accounting for the finite size of the molecules. It was introduced by Johannes Diderik van der Waals in his 1873 doctoral thesis and provided the first semi-quantitative theory describing the gas–liquid change of state and the origin of a critical temperature. The Van der Waals equation is mathematically simple, but it nevertheless predicts the experimentally observed transition between vapor and liquid, which is not possible by the ideal gas model.

ZERO GROWTH LINE = NULLCLINE = ISOCLINE

www.ingramcontent.com/pod-product-compliance
Ingram Content Group UK Ltd.
Pitfield, Milton Keynes, MK11 3LW, UK
UKHW052039301026
9729IPUK00019B/309

9 780750 339322